Irrigation - Water Productivity and Operation, Sustainability and Climate Change

Edited by Sandra Ricart,
Antonio M. Rico and Jorge Olcina

Published in London, United Kingdom

IntechOpen

Supporting open minds since 2005

Irrigation - Water Productivity and Operation, Sustainability and Climate Change
http://dx.doi.org/10.5772/intechopen.77413
Edited by Sandra Ricart, Antonio M. Rico and Jorge Olcina

Contributors
Luis Gurovich, Fernando Riveros, Qudrat Ullah Khan, Obaid Ullah Sayal, Abdul Qadir, Muhammad Irfan, Habib Ali, Nadia Jamil, Sajid Rasheed Ahmad, Hani Mansour, Willians Ribeiro Mendes, Salah Er-Raki, Fábio Meneghetti Ugulino De Araújo, Sandra Ricart

First published in London, United Kingdom, 2019 by IntechOpen
IntechOpen is the global imprint of INTECHOPEN LIMITED, registered in England and Wales, registration number: 11086078, 7th floor, 10 Lower Thames Street, London, EC3R 6AF, United Kingdom
Printed in Croatia

British Library Cataloguing-in-Publication Data
A catalogue record for this book is available from the British Library

Additional hard and PDF copies can be obtained from orders@intechopen.com

Irrigation - Water Productivity and Operation, Sustainability and Climate Change
Edited by Sandra Ricart, Antonio M. Rico and Jorge Olcina
p. cm.
Print ISBN 978-1-78984-676-8
Online ISBN 978-1-78984-677-5
eBook (PDF) ISBN 978-1-78985-304-9

For EU product safety concerns:
IN TECH d.o.o., Prolaz Marije Krucifikse Kozulić 3, 51000 Rijeka, Croatia, info@intechopen.com or visit our website at intechopen.com.

Meet the editors

Sandra Ricart is a geographer with a PhD in Experimental Sciences and Sustainability (2014, University of Girona, Spain). Awarded by different national research institutions, she carried out her predoctoral research at different universities in France and Italy. As a postdoctoral researcher for the "Water and Territory" research group at the University of Alicante (Spain), she participates in national and European research projects, and she has published chapter books and about 20 manuscripts in JCR and Scopus indexed journals focused on irrigation systems, water governance, ecosystem services, un-conventional water resources, and climate change adaptation. She has been an Assistant Professor in Masters in Planning and Management of Natural Risks at University of Alicante since 2018.

Antonio M. Rico is a Professor of Regional Geographical Analysis at the University of Alicante (Spain). He received his MSc degree from the University of Alicante in Territorial Planning in 1989, and his PhD in Geography in 1993. He is the Director of the Interuniversity Institute of Geography of the University of Alicante and the Director of the journal *Investigaciones Geográficas* (indexed in Scopus). He is the Principal Investigator of the "Water and Territory" research group (University of Alicante). With more than a hundred publications in indexed journals and edited books, his research focuses on issues related to agricultural and urban water uses, overexploitation of aquifers, operation of irrigation structures, climate risks (droughts and floods), spatial planning, and un-conventional water resources.

Jorge Olcina is a Professor of Regional Geographical Analysis at the University of Alicante (Spain). He received his PhD in Geography in 1993 from the same university. He has been a visiting professor at universities in Spain, Europe, and Latin America. With near to two hundred publications in indexed journals and books, his research has focused on issues related to climate change mitigation and adaptation measures and climate risks management. He has been a Principal Investigator of the Competitive Research Group of the University of Alicante in "Climate and Territorial Planning" since 2002; and a member of the research teams "History and Climate" and "Water and Territory". He has been the President of the Association of Spanish Geographers (AGE) since November 2017.

Contents

Preface **XI**

Chapter 1 **1**
Introductory Chapter: Addressing Past Claims and Oncoming Challenges for Irrigation Systems
by Sandra Ricart, Jorge Olcina and Antonio M. Rico

Chapter 2 **11**
Agronomic Operation and Maintenance of Field Irrigation Systems
by Luis A. Gurovich and Luis Fernando Riveros

Chapter 3 **27**
Integrating Remote Sensing Data into Fuzzy Control System for Variable Rate Irrigation Estimates
by Willians Ribeiro Mendes, Fábio Meneghetti U. Araújo and Salah Er-Raki

Chapter 4 **57**
Performance of Water Desalination and Modern Irrigation Systems for Improving Water Productivity
by Hani Abdelghani Mansour, Ren Hongjouan, Hu Jiandong, Bao Hong Feng and Liang Changmei

Chapter 5 **81**
Vulnerability of Environmental Resources in Indus Basin after the Development of Irrigation System
by Muhammad Irfan, Abdul Qadir, Habib Ali, Nadia Jamil and Sajid Rashid Ahmad

Chapter 6 **101**
Spate Irrigation: Impact of Climate Change with Specific Reference to Pakistan
by Qudrat Ullah Khan and Obaid Ullah Sayal

Preface

Choices for irrigation system development and management include a large range of technical, operational, economic, and social factors. Irrigation, as a complex socio-ecological system, deals with both the uncertainty of human-nature nexus dynamics and the interdependencies resulting from climate change. Irrigation systems have been under pressure to produce more with lower supplies of water. Globally, irrigation was by far the largest water consumer with between 90–94% of global water consumption. In addition, agriculture is the sector most affected by water scarcity, as it accounts for 70% of global freshwater use. Debates over irrigation system management in the Anthropocene have increasingly been framed in relation to social, economic, and environmental impacts and benefits, stimulating policy framework changes at different scales. In order to ensure irrigation system maintenance and development, technical innovation and social approaches should be understood and analyzed as complementary, as it happens in the water-energy-food nexus. This integrated approach addresses the different gaps by: (a) promoting water and food integrated approaches; (b) improving water efficiency and management at plot scale; (c) ensuring sustainable management of natural ecosystems; and (d) adapting irrigation systems to face water scarcity and environmental risks under climate change scenarios. This book, entitled *Irrigation – Water productivity and operation, sustainability and climate change*, aims to provide examples of consistent progress on mechanisms and approaches related to irrigation system challenges and gaps, such as water productivity, alternative water sources, environmental impacts, and climate change. This collection emphasizes the relevance of innovation and case study analysis to improve knowledge on some of the benefits and limitations of irrigation systems at different geographical contexts.

The first chapter, entitled "*Agronomic operation and maintenance of field irrigation systems*" presented by Luis A. Gurovich and Luis F. Riveros, starts from the consideration that field irrigation system projects are generally adequately designed and installed, considering soil, climate and crop characteristics, with theoretical high water application and distribution efficiencies. However, in most projects, the actual operation and maintenance strategies do not accurately include these characteristics, resulting in excessive water depths applied, generally exceeding crop water needs, unnecessary energy costs, as well as constraints on reaching potential crop yields and marketable crop quality. To address this gap, this chapter describes an approach to dynamic integration of soil hydrodynamic characteristics, potential evapotranspiration, and crop leaf area index evolution throughout the irrigation season, oriented to integrate smart water management strategies and techniques in the operation and maintenance of farm irrigation systems. In addition, this chapter presents how this dynamic integrative platform has been used by farming companies producing table grapes, wine grapes, and avocados in Perú and México. In line with this first chapter, Willians Riberiro Mendes, Fábio Meneghetti U. Araújo and Salah Er-Raki present the chapter "*Integrating remote sensing data into fuzzy control system for variable rate irrigation estimates*" to discuss the necessity of developing precise management zones to apply efficient variable rate irrigation technologies. The authors propose the use of an intelligent fuzzy inference system based on precision irrigation knowledge, for example, by creating perspective maps to control the rotation speed of the central pivot. Results indicate that data from the edaphoclimatic variables, when well fitted to the fuzzy logic, can solve uncertainties and non-linear behavior of an irrigation system and establish a control model for high precision irrigation. A main benefit of this technology is that, because remote sensing provides quick measurements and easy access to crop information for large irrigation areas, images will be used as inputs. Furthermore, the ability of fuzzy systems to deal with complex systems can help farmers to make better decisions in agricultural processes.

Another mechanism to ensure operation and maintenance of irrigation systems was the search for new water sources (most of them conceived as un-conventional water resources) after most conventional water resources are overexploited or contaminated. An example of this commitment to new water sources can be found in the chapter presented by Hani Abdelghani Mansour in collaboration with Ren Hongjouan, Hu Jiandong, Bao Hong Feng, and Liang Changmei, and entitled "*Performance of water desalination and modern irrigation systems for improving water productivity*". This chapter provides a brief update of the Egyptian water strategy for developing irrigation systems. This strategy is based on the promotion of alternative water sources (such as saline water) and, furthermore, the development of varieties of some traditional crops that are saline resistant by using genetic engineering through which saline-tolerant genes are added to the plant. Through field experiments conducted in Saudi Arabia to analyze the effect of different drip irrigation systems and different saline water concentrations on wheat grain yield, water productivity and ecological effects of using saline water are discussed.

In addition to ensuring irrigation system management and water productivity, environmental risks and climate change issues are equally disruptive to the semi-arid and arid regions on a short and medium term. The chapter presented by Muhammad Irfan, Abdul Qadir, Habib Ali, Nadia Jamil, and Sajid Rashid Ahmad and entitled "*Vulnerability of environmental resources in Indus Basin after the development of irrigation system*" informs the reader about both questions. This chapter recognizes the climatic and topographic characteristics of the Indus Basin, which provides an excellent example for the development of an irrigation system. However, in the race of extensive water use, the environmental resources of the Indus Basin have been compromised after 150 years of developing irrigation systems through the construction of dams, barrages, and canals to divert the maximum river water for irrigation. Consequently, water quality was degraded due to the addition of fertilizers, pesticides, chemicals, municipal sewage, and industrial effluents. To overcome this gap, the authors claim that to ensure ecological requirements at water basin scale and to address natural risks, one must promote water governance through key stakeholders. In order to do that, a review of the Indus Water Treaty is promoted with the aim to meet the environmental issues of changing climate and rising water tensions between India and Pakistan.

Closing the book, the last chapter entitled "*Spate irrigation: Impact of climate change with specific reference to Pakistan*" and presented by Qudrat Ullah Khan and Obaid Ullah Sayal, describes the benefits and limitations of spate irrigation under the changing climate and how the promotion of storage dams could affect the hydrological system of the area and the irrigation practices. For example, the management of floodwater and perennial water has affected the water rights of the community and has changed the cropping pattern and land use, which was previously kept fallow and was now used for cultivation. Climate change can only exacerbate this situation. In fact, climate change has greatly influenced Pakistan in frequent spells of extreme weather events, i.e. floods, glacial lake outbursts, droughts, and heat waves. These extreme events have made the country more vulnerable to climate change, as exemplified in the heavy flood of 2010 (causing damage of approximately $10 billion and flooding 38,600 km^2 – mostly irrigated). In addition, the growth of crops is highly affected by the amount of water and changes in temperature; according to an estimate by year 2040, with the increase in temperature of 0.5–2.5°C, the productivity of the crop will decline by 8–10%.

The main challenge faced by irrigation systems is to produce enough food for a continued increase in population, although water and land competition, water quality standards, and water governance in a changing climate are still driving factors at the regional and global scale. This book covers subjects that are often found with both technical and social approaches, without mutual understanding of how irrigation systems should be perceived as socio-ecological systems. These types of contributions are mandatory in the Anthropocene context as a framework for promoting irrigation system management and governance from an agricultural and social sciences collaboration. Taking into account the background of the authors, this book is primarily addressed to the agricultural research community as it provides knowledge about the value of applying specific mechanisms of water productivity, management, and governance. Furthermore, this book should be useful for authorities and irrigators' communities and associations as a first step towards customizing their interventions at local and regional scales to address economic and environmental issues affecting irrigation systems under a changing climate.

The editors of this book want to thank to Anja Filipovic (as Commissioning Editor) and Marijana Francetic (Author Service Manager) for their kind support and invaluable guidance during the editing process. Our most sincere thanks go to those who have been interested in the book and to all authors who have shared their work and experience as part of this publication.

Sandra Ricart, Antonio M. Rico and Jorge Olcina
University of Alicante,
Spain

Chapter 1

Introductory Chapter: Addressing Past Claims and Oncoming Challenges for Irrigation Systems

Sandra Ricart, Jorge Olcina and Antonio M. Rico

1. Introduction

Water-agriculture nexus is context dependent (water availability and water use depend on spatial and temporal issues), socially constructed (multiple stakeholders' perceptions and interests interact), and technically uncertain (benefits from new technologies are difficult to be estimated and duly evaluated). This means that irrigation systems should be analyzed as hydrosocial cycles [1], which likewise takes into account all of these issues including how water management and water governance are conceived and how climate change impacts could be addressed through a "nexus" approach [2]. In few words, irrigation systems are under pressure to produce more food with lower supplies of water [3]. According to this, water availability and water consumption [4], food productivity and food security [5], environmental awareness [6], population growth [7], rural development [8], and climate change [9] are issues to be considered when irrigation systems are promoted, developed, and managed both globally and locally.

2. Irrigation water consumption: calling for concerted effort

Globally, irrigation was by far the largest water consumer with a share ranging over time about 90% of global water consumption [10]. In addition, agriculture is the sector most affected by water scarcity, as it accounts for 70% of global freshwater withdrawals [11]. In fact, agriculture is both a cause and victim of water scarcity, as the excessive use and degradation of water resources is threatening the sustainability of livelihoods dependent on water and agriculture [12]. Furthermore, as the largest water user globally and a major source of water pollution, agriculture plays a key role in tackling the looming water crises. What can agriculture do to address water scarcity in the context of climate change, while ensuring food and nutrition security? What can irrigation offer to alleviate the impacts and reduce the risks of water scarcity? Both questions have been directly addressed through the achievement of the 2030 Agenda for Sustainable Development and the promotion of Sustainable Development Goals (SDG) [13]. These include the adoption of SDG-6 (*"Ensure availability and sustainable management of water and sanitation for all"*) and SDG-2 (*"End hunger, achieve food security and improved nutrition, and promote sustainable agriculture"*). Both goals are an opportunity to be engaged with key water-scarce countries to inform and orient national policies toward effective, sustainable models, and technologies of water management and food security [14]. Furthermore, both are in line with the Paris

IntechOpen

Agreement of the United Nations Framework Convention on Climate Change (UNFCCC)—entered into force on 2016 with the aim of, among others, recognizing the fundamental priority of safeguarding food security and ending hunger and reducing the particular vulnerabilities of food production systems to the adverse impacts of climate change. Furthermore, the Paris Agreement promotes better resilience of socioeconomic and ecological systems through economic diversification and sustainable management of natural resources [15].

3. Irrigation operation: the need for being climate smart

Observed climate change impacts are already affecting food security through increasing temperatures, changing precipitation patterns, and greater frequency of some extreme events [16]. Increasing temperatures are affecting agricultural productivity in higher latitudes, raising yields of some crops (maize, cotton, and wheat), while yields of others are declining in lower-latitude regions [17]. Changes in land use and an increasing demand for water resources have affected the capacity of ecosystems to sustain food production, ensure freshwater resources supply, provide ecosystem services, and promote rural multifunctionality [18]. According to the special report "*Climate Change and Land: an IPCC special report on climate change, desertification, land degradation, sustainable land management, food security, and governance gas fluxes in terrestrial ecosystems*"—recently published by the Intergovernmental Panel on Climate Change (IPCC)—agriculture, forestry, and other land use (AFOLU) activities accounted for 23% of total net anthropogenic greenhouse gas emissions (GHGs) by the period 2007–2016. However, agriculture is not only a contributor to climate change, it will also be severely affected by climate change [19]. Moreover, some effects of warming on crop yields, increased pest occurrences, and the effects of extreme events (e.g., floods, storms, and droughts) on agricultural production are already observed [20]. Although farmers have long adapted to environmental conditions, the severity of the predicted climate changes may be beyond many farmers' current ability to adapt and improve their agricultural production systems and livelihoods [21]. While increased food production will have to be done in the face of a changing climate and climate variability [22], agricultural and irrigation systems should reduce their carbon cost and its contribution to GHG [23]. In order to address this gap, increasing interest has been focused on ensuring that both agriculture and irrigation become climate smart as a driven factor to ensure food security, improve rural livelihoods, and alleviate environmental risks for small-scale farmers [24]. The multi-dimensional aspects of agricultural production under climate change are captured by the climate-smart agriculture (CSA), an approach in which agriculture is transformed and reoriented under the projected scenarios of climate change [25]. The CSA has three concurrent objectives: (i) sustainably increasing farm productivity and income, (ii) increasing adaptive capacity to climate change, and (iii) reducing GHG emissions [26]. In fact, CSA seeks to enhance productivity, water conservation, livelihoods, biodiversity, resilience to climate stress, and environmental quality [27]. Despite the recognized importance of CSA by the Global Alliance for Climate Smart Agriculture (GASCA) and a range of international and national initiatives focused on climate-smart technologies (CST), the dissemination and uptake of climate smart technologies, tools, and practices is still largely an ongoing, challenging process [28]. At this point, some questions should be addressed:

- To what extent is irrigation an enabler of other CSA technologies and under what conditions (soil/market/demography/crop/water management, etc.)?

- Which type of irrigation technology is more climate resilient to extremes and long-term change (watershed management, small-scale pumping, small reservoirs, etc.)?

- Who benefits and what are the implications for food security and food sovereignty if irrigation becomes an integral part of CSA technologies?

According to the FAO-IPCC Expert Meeting on "Climate Change, Land Use and Food Security" celebrated in 2017 [29], to secure a resilient food system under climate change requires a range of appropriate sustainability metrics to better support integrated and multidisciplinary scenario analyses combining socio-economic and ecological dimensions. Among other measures, experts highlighted (1) the need to integrate technical and economic assessments when measuring the impact of improved water use efficiency (maximizing "crop per drop") vs sustainable water use (optimized renewable use of water within a river basin) and (2) the promotion of participatory research to develop frameworks to manage water, land, agroforestry, and crops under different water demand, supply, and pricing conditions.

4. Irrigation impacts and risks: fixing the environmental limits

According to the Organization for Economic Co-operation and Development (OECD), a key challenge for the agriculture sector is to feed an increasing global population, while at the same time reducing the environmental impact and preserving natural resources for future generations. Agriculture can have significant impacts on the environment [30]. While negative impacts are serious and can include pollution and degradation of soil, water, and air [31], agriculture can also positively affect the environment, for instance by trapping GHG within crops and soils [32], or mitigating flood risks through the adoption of certain farming practices [33]. In recent years, there have been some encouraging signs that the agriculture sector and irrigation activities are capable of meeting its environmental challenges. In particular, farmers have made improvements in the use and management of nutrients [34], pesticides [35], energy [36], and water [37], using less of these inputs per unit of land and adopting more environmentally beneficial practices, such as conservation tillage [38] or soil nutrient testing [39]. Taking into account the urgent challenge of matching demand for food for a larger population using the same land footprint, the Global Water Forum (an initiative of the UNESCO Chair in Water Economics and Transboundary Water Governance) discussed the expansion of irrigated areas and their affection to agroecosystems and sustainability [40]. To mitigate that risk while responding to increased global water needs, agricultural management options could include blending different qualities of water sources [41], matching irrigation methods or promoting deficit irrigation [42], and selecting salt tolerant crops [43]. Whatever methods and strategies are used to increase food production, they must also preserve soil ecological functionality and minimize environmental risks.

5. Irrigation adaptation: water management and alternative water sources

As freshwater resources are under increasing stress in several world regions, with a mismatch between availability and demand and temporal and geographical scales [44], new approaches have been promoted in order to guarantee the agricultural activity (by considering social and economic issues) and irrigation sustainability (by addressing environmental issues) in an integrated way. The first approach is focused on putting more attention to understanding current water management and promoting transition to more adaptive water regimes that take into account environmental, technological, economic, institutional, and cultural characteristics of river basins. This implies a paradigm shift in water management from a prediction and control to a management as a social-learning approach [45]. The second approach has been focused on water availability. That is, the general decreasing trend in water availability and the need for sustainable use of available water resources have led regional and national governments worldwide to seek alternative water sources [46], putting special attention to wastewater reuse and water desalination. The first one is not a "new" water source, but rather a way to waste able to be used for a new water demand. It differs to increase water supply measures such as seawater desalination, which in effect includes a new input to the water cycle [47]. Both concepts, water reuse and seawater desalination, are limited by different key barriers. The first barrier is that their management is more complex than the management of conventional water resources, but also their cost is more expensive than the cost of "environmental" water sources—rivers—due to its conveyance, storage, and distribution in dedicated network infrastructure [48]. The second barrier is that both the public and farmers negatively percept alternative water sources by highlighting their environmental and health risks instead of their benefits (especially in the case of wastewater resources) [49–52]. Furthermore, although there are rules and regulations clearly focused on ensuring standards on food security, yuck factor currently justify the negative to use alternative water resources [53]. It should be noted that addressing the last two barriers are not solely related to technical issues, but to social issues. According to this and irrespective of scientific and engineering based considerations, farmers' opposition and public rejection has the potential to cause water reuse and water desalination projects to fail, before, during, or after their execution [54]. In fact, reuse and desalinated water schemes may face public opposition resulting from a combination of prejudiced beliefs, fear, attitudes, lack of knowledge, and general distrust, which, on the whole, is often not unjustified, judging by the frequent (and highly publicized) failures of wastewater treatment facilities worldwide.

6. Irrigation challenge: welcome to the Anthropocene

The need for capturing, storing, cleaning, and redirecting freshwater resources in efforts to increase water availability even with irregular river flows and unpredictable rainfall has been one of the main challenges of humanity [55]. Resulting impacts on water productivity and security schemes (which requires waterworks from storage and distribution such as dams, pipelines, canals, and water transfers) [56] means that the water cycle has been increasingly controlled by human activities and this was the hallmark of the new geological epoch called the "Anthropocene" [57]. This term is currently used (and discussed) to encompass different geological, ecological, sociological, and behavioral dynamics in recent earth history. The origins of the concept, its terminology, and its socio-political implications have also been widely discussed across the scientific community [58]. In fact, for some authors, the commitment to define a new geological period responds to the *hydrocentric* approach that emerged over the past two decades [59, 60], which focused on managing water resources as a natural water environment duly protected. Some evidences suggest, however, that what are needed are rather *hydrosupportive* approaches in which water management is performed to achieve social goals, which may include, among other factors, the ability to sustain environmental functions [61]. The concept, popularized by the Dutch atmospheric chemist and Nobel Prize-winning Paul Crutzen, is defined to describe a new geologic era caused by the drastic effect of human action on the earth. Taking into account the transdisciplinary nature of the concept, the analysis of human-water interactions requires the collaboration between natural sciences and the humanities, which must simultaneously explore the geophysical, social, and economic forces that shape an increasingly human dominated global hydrologic (and hydrosocial) system [62].

According to the report "*Adapt Now: A global call for leadership on climate resilience*" published on 2019 by the Global Commission on Adaptation, adapting the planet's water resources and systems to the Anthropocene and the new climate reality is a formidable task. Furthermore, it is the main opportunity to improve ecosystems management, grow eco-friendly economies, boost agricultural efficiencies, and planning for natural risks (floods and droughts) from nature-based solutions [63]. In fact, 10 years ago, a report from the Food and Agriculture Organization of the United Nations (FAO) untitled "*Climate change, water and food security*" clearly promoted the applicability of different adaptation measures that deal with climate variability and build upon improved land and water management practices. These measures imply a good understanding of the impact of climate change on available water resources and on agricultural systems, and a set of policy choices, and investments and managerial changes to address them. Some year later and in order to respond to water-food nexus challenges in a coordinated and effective manner, the FAO has developed the *Global Framework for Action to Cope with Water Scarcity in Agriculture in the Context of Climate Change*. It calls for urgent action to cope with water scarcity in agriculture in the context of climate change and growing competition for water resources. The *Global Framework for Action* recognizes the intricate links between climate change, water scarcity, sustainable agriculture, and food security and the importance of addressing these holistically. Its objective is to strengthen the capacity to adapt agriculture to the impacts of climate change and water scarcity and thereby to reduce water-related constraints to achieving the food security and sustainable development goals. This framework is based on the premise that a sustainable pathway to food security in the context of water scarcity lies in maximizing benefits that cut across multiple dimensions of the food-water-climate nexus [64]. This means enabling sustainable agricultural production while reducing vulnerability to water scarcity and optimizing the climate change adaptation and mitigation benefits [65].

Taking into account both the adaptation capacity of irrigation systems from its socio-ecological nature and the requirements for addressing oncoming climate challenges, this book is the first attempt at bringing several fields together to analyze irrigation by combining technical, social, and management approaches.

Author details

Sandra Ricart*, Jorge Olcina and Antonio M. Rico
University of Alicante, Spain

*Address all correspondence to: sandra.ricart@ua.es

References

[1] Linton J, Budds J. The hydrosocial cycle: Defining and mobilizing a relational-dialectical approach to water. Geoforum. 2014;**57**:170-180. DOI: 10.1016/j.geoforum.2013.10.008

[2] Endo A, Tsurita I, Burnett K, Orencio P. A review of the current state of research on the water, energy, and food nexus. Journal of Hydrology: Regional Studies. 2017;**11**:20-30. DOI: 10.1016/j.ejrh.2015.11.010

[3] Levidow L, Zaccaria D, Maia R, Vivas E, Todorovic M, Scardigno A. Improving water-efficient irrigation: Prospects and difficulties of innovative practices. Agricultural Water Management. 2014;**146**:84-94. DOI: 10.1016/j.agwat.2014.07.012

[4] du Plessis A. Current and future water availability. In: du Plessis A, editor. Water as an Inescapable Risk. Current Global Water Availability, Quality and Risks with a Specific Focus on South Africa. Cham: Springer International Publishing; 2019. pp. 3-11

[5] Dias J, Reidsma P, Giller K, Todman L, Whitmore A, van Ittersum M. Sustainable development goal 2: Improved targets and indicators for agriculture and food security. Ambio. 2019;**48**(7):685-698. DOI: 10.1007/s13280-018-1101-4

[6] Gulse Bal HS. Karakas G, Environmental education at Faculty of Agriculture and changing awareness, attitude and behaviour towards environment in Turkey. Journal of Agricultural Science and Technology. 2018;**20**:869-882

[7] Brown TC, Mahat V, Ramirez JA. Adaptation to future water shortages in the United States caused by population growth and climate change. Earth's Future. 2019;**7**(3):219-234. DOI: 10.1029/2018EF001091

[8] Sumner J. Contested sustainabilities: The post-carbon future of Agri-food, rural development and sustainable place-making. Journal of Agriculture Food Systems and Community Development. 2019;**8**(4):219-220. DOI: 10.5304/jafscd.2019.084.018

[9] Iglesias A, Garrote L. Adaptation strategies for agricultural water management under climate change in Europe. Agricultural Water Management. 2015;**155**:113-124. DOI: 10.1016/j.agwat.2015.03.014

[10] Zhang X, Liu J, Zhao X, Yang H, Deng X, Jiang X, et al. Linking physical water consumption with virtual water consumption: Methodology, application and implications. Journal of Cleaner Production. 2019;**228**:1206-1217. DOI: 10.1016/j.jclepro.2019.04.297

[11] Huang Z, Hejazi M, Tang Q, Vernon CR, Liu Y, Chen M, et al. Global agricultural green and blue water consumption under future climate and land use changes. Journal of Hydrology. 2019;**574**:242-256. DOI: 10.1016/j.jhydrol.2019.04.046

[12] De Fraiture C, Molden D, Wichelns D. Investing in water for food, ecosystems, and livelihoods: An overview of the comprehensive assessment of water management in agriculture. Agricultural Water Management. 2010;**97**(4):495-501. DOI: 10.1016/j.agwat.2009.08.015

[13] Entezari A, Wang RZ, Zhao S, Mahdinia E, Wang JY, Tu YD, et al. Sustainable agriculture for water-stressed regions by air-water-energy management. Energy. 2019;**181**:1121-1128. DOI: 10.1016/j.energy.2019.06.045

[14] Gil JDB, Reidsma P, Giller K, Todman L, Whitmore A, van Ittersum M. Sustainable development goal 2: Improved targets and indicators

for agriculture and food security. Ambio. 2019;**48**:685-698. DOI: 10.1007/s13280-018-1101-4

[15] Amjath-Babu TS, Aggarwal PK, Vermeulen S. Climate action for food security in South Asia? Analyzing the role of agriculture in nationally determined contributions to the Paris agreement. Climate Policy. 2019;**19**(3):283-298. DOI: 10.1080/14693062.2018.1501329

[16] van Meijl H, Havlik P, Lotze-Campen H, Stehfest E, Witzke P, Pérez Domínguez I, et al. Comparing impacts of climate change and mitigation on global agriculture by 2050. Environmental Research Letters. 2018;**13**(6):064021. DOI: 10.1088/1748-9326/aabdc4

[17] Boonwichai S, Shrestha S, Babel MS, Weesakul S, Datta A. Climate change impacts on irrigation water requirement, crop water productivity and rice yield in the Songkhram River basin, Thailand. Journal of Cleaner Production. 2018;**198**:1157-1164. DOI: 10.1016/j.jclepro.2018.07.146

[18] Ricart S, Olcina J, Rico AM. Evaluating public attitudes and farmers' beliefs towards climate change adaptation: Awareness, perception, and populism at European level. Land. 2019;**8**(1):4. DOI: 10.3390/land8010004

[19] Zimmermann A, Benda J, Webber H, Jafari Y. Trade, Food Security and Climate Change: Conceptual Linkages and Policy Implications. Rome: FAO; 2018. 48p

[20] Ren X, Lu Y, O'Neill BC, Weitzel M. Economic and biophysical impacts on agriculture under 1.5°C and 2°C warming. Environmental Research Letters. 2018;**13**(11):115006. DOI: 10.1088/1748-9326/aae6a9

[21] Hellin J, Fisher E. Climate smart agriculture and non-agricultural livelihood transformation. Climate. 2019;**7**:48. DOI: 10.3390/cli7040048

[22] Kummu M, Gerten D, Heinke J, Konzmann M, Varis O. Climate-driven interannual variability of water scarcity in food production potential: A global analysis. Hydrology and Earth System Sciences. 2014;**18**:447-461. DOI: 10.5194/hess-18-447-2014

[23] Tubiello FN, Salvatore M, Ferrara AF, House J, Federici S, Rossi S, et al. The contribution of agriculture, forestry and other land use activities to global warming, 1990-2012. Global Change Biology. 2015;**21**(7):2655-2660. DOI: 10.1016/j.agsy.2018.09.009

[24] Makate C, Makate M, Mango N, Siziba S. Increasing resilience of smallholder farmers to climate change through multiple adoption of proven climate-smart agriculture innovations. Lessons from southern Africa. Journal of Environmental Management. 2019;**231**:858-868. DOI: 10.1016/j.jenvman.2018.10.069

[25] Lipper L, Thornton P, Campbell BM, Baedeker T, Braimoh A, Bwalya M, et al. Climate-smart agriculture for food security. Natural Climate Change. 2014;**4**:1068-1072. DOI: 10.1038/nclimate2437

[26] Rao NH. Big data and climate smart agriculture–Status and implications for agricultural research and innovation in India. Proceedings of the Indian National Science Academy. 2018;**84**(3):625-640. DOI: 10.16943/ptinsa/2018/49342

[27] Olayide OE, Tetteh IK, Popoola L. Differential impacts of rainfall and irrigation on agricultural production in Nigeria: Any lessons for climate-smart agriculture? Agricultural Water Management. 2016;**178**:30-36. DOI: 10.1016/j.agwat.2016.08.034

[28] Janssen SJC, Porter CH, Moore AD, Athanasiadis JN, Foster I, Jones JW,

et al. Towards a new generation of agricultural system data, models and knowledge products: Information and communication technology. Agricultural Systems. 2017;**155**:269-288. DOI: 10.1016/j.agsy.2016.09.017

[29] Elbehri A, Challinor A, Verchot L, Angelsen A, Hess T, Ouled Belgacem A, et al. FAO-IPCC Expert Meeting on Climate Change, Land Use and Food Security: Final Meeting Report. Rome: FAO HQ; 2017

[30] Popescu C, Dragomir L, Popescu G, Horablaga A, Chis C. Evaluation of the impact of agriculture on the environment in EU27 countries with cluster analysis. Journal of Biotechnology. 2016;**231**:S103. DOI: 10.1016/j.jbiotec.2016.05.361

[31] Smith L, Inman A, Lai X, Zhang H, Fanqiao M, Jianbin Z, et al. Mitigation of diffuse water pollution from agriculture in England and China, and the scope for policy transfer. Land Use Policy. 2017;**61**:208-219. DOI: 10.1016/j.landusepol.2016.09.028

[32] Frank S, Havlík P, Soussana J-F, Levesque A, Valin H, Wollenberg E, et al. Reducing greenhouse gas emissions in agriculture without compromising food security? Environmental Research Letters. 2017;**12**:105004. DOI: 10.1088/1748-9326/aa8c83

[33] Ullah W, Nafees M, Khurshid M, Nihei T. Assessing farmers' perspectives on climate change for effective farm-level adaptation measures in Khyber Pahtunkhwa, Pakistan. Environmental Monitoring and Assessment. 2019;**191**:547. DOI: 10.1007/s10661-019-7651-5

[34] Denny RHC, Marquart-Pyatt ST, Houser M. Understanding the past and present and predicting the future: Farmers' use of multiple nutrient best management practices in the upper midwest. Society and Natural Resources. 2019;**32**(7):807-826. DOI: 10.1080/08941920.2019.1574045

[35] Bagheri A, Emami N, Damalas CA, Allahyari MS. Farmers' knowledge, attitudes, and perceptions of pesticide use in apple farms of northern Iran: Impact on safety behaviour. Environmental Science and Pollution Research. 2019;**26**(9):9343-9351. DOI: 10.1007/s11356-019-04330-y

[36] He R, Jin JJ, Gong HZ, Tian YH. The role of risk preferences and loss aversion in farmers' energy-efficient appliance use behaviour. Journal of Cleaner Production. 2019;**215**:305-314. DOI: 10.1016/j.jclepro.2019.01.076

[37] Garg KK, Wani SP, Patil MD. A simple and farmer-friendly decision support system for enhancing water use efficiency in agriculture: Tool development, testing and validation. Current Science. 2016;**110**(9):1716-1729. DOI: 10.18520/cs/v110/i9/1716-1729

[38] Bossange AV, Knudson KM, Shrestha A, Harben R, Mitchell JP. The potential for conservation tillage adoption in the San Joaquin Valley, California: A qualitative study of farmer perspectives and opportunities for extension. PLoS One. 2016;**11**(12):e0167612. DOI: 10.1371/journal.pone.0167612

[39] Alobwede E, Leake JR, Pandhal J. Circular economy fertilization: Testing micro and macro algal species as soil improvers and nutrient sources for crop production in greenhouse and field conditions. Geoderma. 2019;**334**:113-123. DOI: 10.1016/j.geoderma.2018.07.049

[40] Steidl J, Schuler J, Schubert U, Dietrich O, Zander P. Expansion of an existing water management model for the analysis of opportunities and impacts of agricultural irrigation under climate change conditions. Water. 2015;**7**(11):6351-6377. DOI: 10.3390/w7116351

[41] Malakar A, Snow DD, Ray C. Irrigation water quality–A contemporary perspective. Water. 2019;**11**(7):1482. DOI: 10.3390/w11071482

[42] Al-Ghobari HM, Dewidar AZ. Deficit irrigation and irrigation methods as on-farm strategies to maximize crop water productivity in dry areas. Journal of Water and Climate Change. 2018;**9**(2):399-409. DOI: 10.2166/wcc.2017.014

[43] Sobhanian H, Aghaei K, Komatsu S. Changes in the plant proteome resulting from salt stress: Toward the creation of salt-tolerant crops? Journal of Proteomics. 2011;**74**(8):1323-1337. DOI: 10.1016/j.jprot.2011.03.08

[44] Chen B, Han MY, Peng K, Zhou SL, Shao L, Wu XF, et al. Global land-water nexus: Agricultural land and freshwater use embodied in worldwide supply chains. The Science of the Total Environment. 2018;**613-614**:931-943. DOI: 10.1016/j.scitotenv.2017.09.138

[45] Iglesias A, Garrote L. Adaptation strategies for agricultural water management under climate change in Europe. Agricultural Water Management. 2015;**155**:113-124. DOI: 10.1016/j.agwat.2015.03.014

[46] Giannoccaro G, Arborea S, de Gennaro BC, Iacobellis V, Piccinni AF. Assessing reclaimed urban wastewater for reuse in agriculture: Technical and economic concerns for Mediterranean regions. Water. 2019;**11**:1511. DOI: 10.3390/w11071511

[47] Navarro T. Water reuse and desalination in Spain–Challenges and opportunities. Journal of Water Reuse and Desalination. 2018;**8**(2):153-168. DOI: 10.2166/wrd.2018.043

[48] Voutchkov N. Desalination Project Cost Estimating and Management. Boca Raton (Florida): Taylor & Francis Group; 2019

[49] Voulvoulis N. Water reuse from a circular economy perspective and potential risks from an unregulated approach. Current Opinion in Environmental Science & Health. 2018;**2**:32-45. DOI: 10.1016/j.coesh.2018.01.005

[50] Smith HM, Brouwer P, Jeffrey P, Frijns J. Public responses to water reuse–Understanding the evidence. Journal of Environmental Management. 2018;**207**:43-50. DOI: 10.1016/j.jenvman.2017.11.021

[51] Antwi-Agyei P, Peasey A, Biran A, Bruce J, Ensink J. Risk perceptions of wastewater use for urban agriculture in Accra, Ghana. PLoS One. 2016;**11**(3):e0150603. DOI: 10.1371/journal.pone.0150603

[52] Padilla-Rivera A, Morgan-Sagastume JM, Noyola A, Güereca LP. Addressing social aspects associated with wastewater treatment facilities. Environmental Impact Assessment. 2016;**57**:101-113. DOI: 10.1016/j.eiar.2015.11.007

[53] Hanjra MA, Wichelns D, Drechsel P. Investing in water management in rural and urban landscapes to achieve and sustain global food security. In: Zeunert J, Waterman T, editors. Routledge Handbook of Landscape and Food. New York: Routledge; 2018. pp. 278-296

[54] Ricart S. Challenges on European irrigation governance: From alternative water resources to key stakeholders' involvement. Journal of Ecology & Natural Resources. 2019;**3**(2):000161. DOI: 10.23880/jenr-16000161

[55] Seelen LMS, Flaim G, Jennings E, de Senerpont Domis LN. Saving water for the future: Public awareness of water usage and water quality. Journal of Environmental Management. 2019;**242**:246-257. DOI: 10.1016/j.jenvman.2019.04.047

[56] Al-Kalbani MS, Price MF, O'Higgins T, Ahmed M, Abahussain A. Integrated environmental assessment to explore water resources management in Al Jabal Al Akhdar, Sultanate of Oman. Regional Environmental Change. 2016;**16**(5):1345-1361. DOI: 10.1007/s10113-015-0864-4

[57] Vorosmarty CJ, Meybeck M, Pastore CL. Impair-then-repair: A brief history & global-scale hypothesis regarding human-water interactions. Daedalus. 2015;**144**(3):94-109

[58] Malm A, Hornborg A. The geology of mankind? A critique of the Anthropocene narrative. Anthropocene Review. 2014;**1**(1):62-69. DOI: 10.1177/2053019613516291

[59] Arnell NW, van Vuuren DP, Isaac M. The implications of climate policy for the impacts of climate change on global water resources. Global Environmental Change. 2011;**21**(2):592-603

[60] Ates S, Isik S, Keles G, Aktas AH, Louhaichi M, et al. Evaluation of deficit irrigation for efficient sheep production from permanent sown pastures in a dry continental climate. Agricultural Water Management. 2013;**119**:135-143. DOI: 10.1016/j.agwat.2012.12.017

[61] Muller M. Sustainable water management in the Anthropocene. Proceedings of the Institution of Civil Engineers. 2017;**170**(4):187-195. DOI: 10.1680/jensu.15.00028

[62] Waters CN, Zalasiewicz J, Summerhayes C, Barnosky AD, Poirier C, al e. The Anthropocene is functionally and stratigraphically distinct from the Holocene. Science. 2016;**351**(6269):aad2622

[63] Boelee E, Janse J, Le Gal A, Kok M, Alkemade R, Ligtvoet W. Overcoming water challenges through nature-based solutions. Water Policy. 2017;**19**(5):820-836. DOI: 10.2166/wp.2017.105

[64] Khacheba R, Cherfaoui M, Hartani T, Drouiche N. The nexus approach to water-energy-food security: An option for adaptation to climate change in Algeria. Desalination and Water Treatment. 2018;**131**:30-33. DOI: 10.5004/dwt.2018.22950

[65] Brouziyne Y, Abouabdillah A, Hirich A, Bouabid R, Zaaboul R, Benaabidate L. Modeling sustainable adaptation strategies toward climate-smart agriculture in a Mediterranean watershed under projected climate change scenarios. Agricultural Systems. 2018;**162**:154-163. DOI: 10.1016/j.agsy.2018.01.024

Chapter 2

Agronomic Operation and Maintenance of Field Irrigation Systems

Luis A. Gurovich and Luis Fernando Riveros

Abstract

Worldwide experience indicates that projected economic returns on investments in field irrigation systems are seldom obtained by farmers, due to improper strategies on irrigation scheduling, lack of operational control, and limited feedback on the actual performance of irrigation systems, in terms of application efficiency and uniformity. An approach to dynamic integration of soil hydrodynamic characteristics, potential evapotranspiration, and crop leaf area index evolution throughout the irrigation season is detailed, oriented to integrate smart water management strategies and techniques in the operation and maintenance of farm irrigation systems. This dynamic integrative platform has been used in Perú and México by actual farming companies producing table grapes, wine grapes, avocado, and bell peppers exported to international markets; this chapter documents its practical results in terms of water and energy savings, crop yield, and fruit quality.

Keywords: English, agronomy, smart irrigation management, real-time irrigation scheduling, dynamic crop ET coefficient, soil/water monitoring, farmers' attitudes, irrigation infrastructure modernization, irrigation investments and return optimization

1. Introduction

Achieving an efficient use of natural resources and other production factors is a common goal of many of the current policies aimed at the sustainability of human activity; irrigation of agricultural crops uses about 80% of the total freshwater available for all human activities; thus, improving irrigation efficiency is a main endeavor to provide sustainability to this vital resource availability [1, 2].

Worldwide experience indicates that projected economic returns on investments in field irrigation systems are seldom fully obtained by farmers, due to improper strategies on irrigation scheduling, lack of operational control, and limited feedback on the actual performance of irrigation systems, in terms of application efficiency and uniformity. Field irrigation system projects are generally properly designed and installed, considering soil, climate and crop characteristics, with theoretical high water application and distribution efficiencies. However, in most projects, its actual operation and maintenance strategies do not accurately include these characteristics, resulting in excessive water depths applied, generally well over actual crop water needs, unnecessary energy costs, as well as constraints on reaching potential crop yields and marketable fruit quality. Also, irrigation systems' cumulative deterioration conditions after its installation in the field, due to lack of proper maintenance and timely spare parts replacement, result in a significant reduction of the cost effectiveness of farm investments in irrigation infrastructure [3].

Lack of operational control, limited feedback on the actual performance of irrigation systems, in terms of application efficiency and uniformity, limited use of agrometeorological and crop development data to assess crop water needs, and scant follow-up of soil water content dynamics as an indicator of the fit between actual water applied and actual water evapotranspirated by the crop, as well as limitations on human resources knowledge and training, are the major issues explaining the situation described above.

Irrigation scheduling is related to the farmers' decision process concerning ""when" to irrigate and "how much" water to apply, in order to maximize agriculture production profit. Knowledge on crop water requirements and yield responses to water, as well as specific irrigation equipment constraints, limitations relative to the water supply system, and financial and economic implications of irrigation practice, must be integrated in any rational strategy to optimize pressurized irrigation systems use [4–7].

When appropriate water application techniques (i.e., irrigation system physical characteristics) are correctly coupled with irrigation scheduling (i.e., the volume and timeliness of water applications), as well as the implementation of irrigation system proper maintenance strategies, it is possible to optimize available water for irrigation, achieve potential crop yield/quality, and reduce irrigation costs. Investing resources in an up-to-date technological, sophisticated irrigation system is not by itself enough to attain high levels of performance, if its operation and maintenance are not updated accordingly [8].

Research has made available many tools, including procedures to compute crop water requirements, simulate soil water balance, estimate the impact of water deficits on yield and evaluating the economic returns of irrigation; however, irrigation scheduling and comprehensive irrigation equipment maintenance protocols are not yet utilized by the majority of farmers. Furthermore, only limited irrigation scheduling information is utilized worldwide by irrigation system managers, extensionists, or farmer advisers. It is recognized, however, that the adoption of appropriate irrigation scheduling practices generally leads to increased yield and profit improvements for farmers, significant water and energy savings, reduced environmental impact of irrigation, and long-term sustainability of irrigated agriculture [9–13].

Integration of soil hydrodynamic characteristics, potential evapotranspiration, and crop leaf area index evolution throughout the irrigation season, with actual irrigation operation data, and soil water content periodic measurements, is needed to implement smart water management strategies, aimed to optimize the economic return of investments in irrigation equipment at the farm level, as well as to reduce its operational costs and ensure continuous optimal soil water availability conditions to crops [14].

Pressurized irrigation application equipment (drip, microjet, or microsprinkler) is a high precision machine, which allows the producer to obtain the highest productivity of their agricultural crops, and at the same time, achieve specific quality characteristics, in accordance to market demands. Like any high precision machine, its design, installation, operation, and optimal maintenance are absolutely essential to achieve the objectives of high production and high quality of any viticulture, fruit, or horticultural plantation. If the design, installation, operation and/or maintenance of the systems are not optimal, generally, its negative effects on crop production and quality are more detrimental than the incorrect use of surface irrigation, because root crop soil volume wetted by each emitter (dripper, microjet, or microsprinkler) is restricted, being essential to maintain in this restricted soil volume-specific water and nutrients, salinity, acidity (pH), and oxygen availability conditions, continuously throughout the production season [15].

This chapter reports the main components and actual use of an interactive, dynamic, and relational database management system (RDBMS), an irrigation scheduling platform, using structured query language (SQL) for querying, maintaining, and updating the database [16]. The platform is designed to implement smart water management strategies and techniques in the operation and maintenance of farm irrigation systems in actual plantations, fruit orchards, and vineyards irrigated by drip or microsprinkler systems [6, 11, 17, 18]. The platform allows graphic representation of relevant data and processed results, automatically updating all the information required in any time span and/or in any irrigation sector combination, using interactive, easily understandable dashboards. Specific considerations for field irrigation system maintenance are also discussed in this chapter, with an analysis on the constraints for the platform adoption by farming personnel, farm decision-making stakeholders, and farm advisors.

2. Irrigation scheduling interactive platform

The interactive platform developed integrates soil hydrodynamic characteristics relevant to irrigation scheduling, with crop water requirements, based on atmospheric evaporative demand and the evolution of crop leaf area index throughout the irrigation season, as well as with the irrigation system daily effective operation, in terms of actual water depths (expressed in mm or m^3/hectare, being 1 mm = 10 m^3/hectare) applied to each irrigated sector. Independently, information on the evolution of soil water content is also integrated, allowing next 5 days' irrigation schedules to be automatically modified, aiming to maintain continuous soil water availability conditions to the crop, if the soil profile water content trend is increasing or decreasing with respect to a specific target range [6, 9, 11, 18–20].

2.1 Soil hydrodynamic properties relevant to irrigation scheduling

The platform calculates the soil volume effectively providing water to crop roots, considering soil stratification depths and textures, and the integrated water volume stored at field capacity, calculated using the "Soil Water Characteristics Hydraulic Properties Calculator" [21], assuming that water distribution in the soil below each irrigation emitter forms an ellipsoid, with specific a, b, and c radii measured in soil observation trenches at the onset of the irrigation season [22, 23] (**Figure 1**). We have repeated soil water distribution field observations on a bimonthly basis, and for most soils, a, b, and c values remain fairly constant throughout the irrigation season.

Management of the allowed soil water depletion (MAD) by ETc [8, 9], defined as the percentage of soil water stored at field capacity in the effective soil water volume, is the threshold to initiate the next irrigation event; it considers soil root crop distribution and its water extraction pattern, rootstock relative drought resistance, as well as soil major texture class, crop value, and water costs; this threshold can also be modified according to specific crop phenology stages [19, 20].

The platform is programmed to schedule irrigation based on the "*variable frequency—variable water depth*" approach [4, 6, 7, 9, 13]; however, a maximal irrigation time value for each irrigation cycle is defined for each soil dominant structure, to avoid water percolation in lighter soils and to avoid surface water ponding or partial soil saturation in heavier soils. Thus, during high atmospheric evaporative demand periods, irrigation water depth equivalent to daily ETc in sandy soils determines the need of several watering events or cycles throughout the day, and in clay soils, irrigation is applied in 2–3 days cycle intervals, to replace the total water depth corresponding to Σ (daily ETc since the last irrigation event).

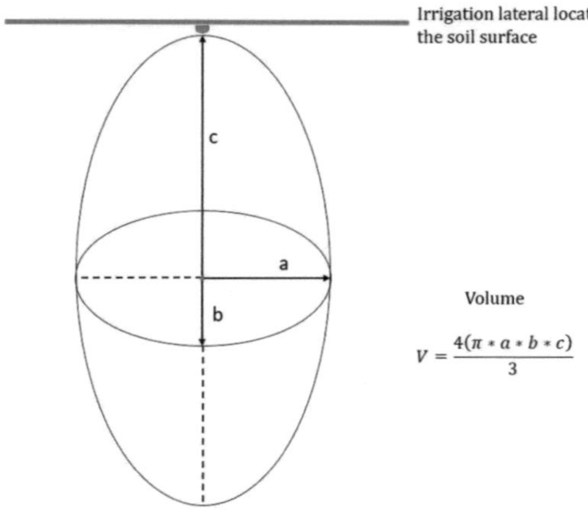

Figure 1.
Ellipsoid representing soil water distribution below a dripper.

2.2 Crop evapotranspiration (ETc) assessment

The platform makes use of the modified FAO Penman-Monteith model [8, 9, 10, 13, 19, 24, 25] to define actual daily crop water use, as the product of site-specific atmospheric evaporative demand maximum value, assuming: (1) unlimited moisture availability and ambient atmospheric conditions, or potential evapotranspiration (ETp) [26] and (2) actual crop leaf area index, expressed as a crop coefficient function (Kc) [27–29].

2.2.1 Potential ET (ETp)

Daily potential ETp data are widely provided by government or private meteorological weather station services in significantly large irrigated areas around the world [8, 14]; additionally, the use of automatic weather stations at the farm is growing rapidly in many countries, because it represents a marginal additional investment in the context of pressurized irrigation systems. Weather stations world nets, like Climwat provided by FAO [30], or regional nets are useful sources of ETp major components' information (air temperature and humidity, solar radiation, and wind direction and intensity). Routine use of ETp data by farmers for irrigation scheduling purposes has not being widely adopted; extensive farm extension work on the subject is urgently needed, especially in areas with restricted water resources.

The representativeness of a single weather station to provide accurate ETp data is highly dependent on topography, crop surrounding areas cultivation pattern, due to albedo effects, as well as on microatmospheric specific conditions. Installation of at least one weather station every 100 hectares is highly recommended; moreover, daily ETp differences in relatively close spots within the irrigated field are highly correlated to one or two climatic parameters (i.e., maximal day temperature, or solar radiation), thus the use of single sensors adequately located, instead of complete weather stations, can be used accordingly [10, 12, 19].

Accurate ETp assessment using weather stations requires keeping adequate maintenance protocols, regarding sensor periodic cleaning and at least a yearly calibration [31]; the extended amount of data provided daily by a specific weather station must be addressed using big data analysis tools and models [16, 20], coupling it with actual data on the irrigation system operation, as well with soil water dynamics in the wetted soil volume, to fully achieve its potential aimed to provide continuous optimal conditions of soil water availability, coupled with water and energy savings.

Incorporating online, real-time weather sensors, with irrigation system sensors data (pressure and discharge, water pH and salinity) and soil water content data, is a useful example of the Internet of Things applied into farming decision-making processes; its rapid adoption by large number of farmers within a specific agriculture area could account for a positive and sound impact in smart water management [5, 11, 19, 20].

2.2.2 Crop ET coefficients [Kc = f(crop phenology)]

Actual water evapotranspirated by a crop (ETc) is not only determined by ETp; an estimation of transpiration canopy is also needed. This estimation corresponds to the concept of leaf area index (IAF; m^2 of transpiring leaves/m^2 of cultivated land) [23, 28, 32]; for irrigation scheduling purposes, this concept is generally expressed as the "crop coefficient" (Kc) [27, 29, 33–35] which in fact is a time function, since IAF varies from bare soils at the end of winter ($Kc_{initial}$ = 0.1–0.15), representing direct soil surface evaporation, up to Kc_{max} = 0.8–1.2), when the maximal IAF is attained. The Kc = (t) function can be represented by a double sigmoid curve (**Figure 2**) for the initial three crop phenology stages (budbreak, flowering, and veraison) [36, 37]; a constant maximal value from veraison to harvest, and a linear decline for the postharvest irrigation stage, reaching a $Kc_{final} = Kc_{initial}$ [37]. The maximal Kc_{max} value has been widely reported for most irrigated crops [27]; at flowering [$Kc_{flower} = (Kc_{max}/Kc_{initial})/2$] [26, 38].

For irrigation scheduling, the Kc daily value is obtained from **Figure 2** or from an equivalent table; the main concern is related to the onset data for each phenology stage, which seldom can be predicted accurately from crop models and needs periodic field observations throughout the irrigation season. Modifying these dates on the platform is a very simple procedure and the Kc = f (t) function is easily recalculated. Different crops and/or different cultivation locations for the same crop can have quite different Kc curve (**Figure 2**) shapes, due to the relative onset date and duration of each phenology stage, but essentially, this schematic representation can be adapted to these differences.

2.3 Automatic Kc value adjustments, based on soil water content data

Data on the evolution of soil water content at specific depths and distances from the irrigation lateral enable the platform to automatically adjust Kc values for the next 5 days, with the aim to increment or reduce recommendations for next irrigation dates and water depths to be applied, in order to keep a constant soil water

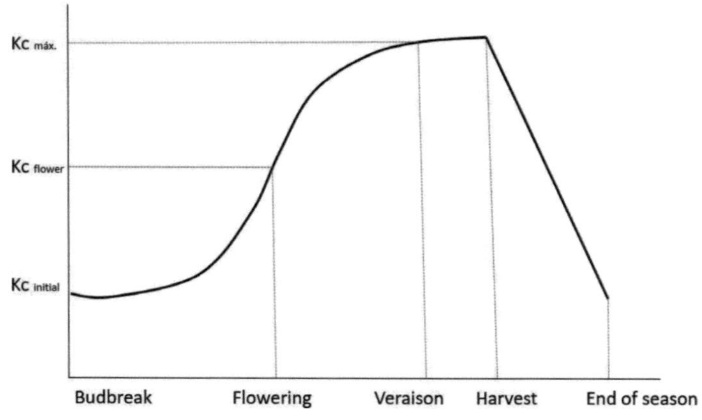

Figure 2.
The Kc = f (phenology stage) function.

availability condition; these data are obtained by using soil water content probes, providing either real-time or periodic measurements with portable soil probes. This platform feature is an independent checking for the balance between calculated ETc and actual depth water applied, enabling to automatically correct eventual errors in the calculated ETc. If the calculated ETc value is lower than the actual ETc, platform recommendations will determine underirrigation and a gradual reduction in the soil water content, while if calculated ETc > actual ETc, overirrigation will determine a gradual increment in the soil water content. Soil water content increments or reductions over 5% between consecutive measurements trigger automatic modifications on Kc values for the next 5 days and thus, the process is self-adjusted. All Kc adjustments are kept in an historical file, to be used as platform input data for the following irrigation seasons (see Section 2.2.2) [35].

Adjustment of Kc daily values related to soil water dynamics, as affected by the balance between calculated ETc and actual irrigation water depth applied, represents an automatic fine-tune procedure on irrigation scheduling, aimed to keep a constant crop water availability condition, simultaneously considering atmospheric evaporative demand, crop IAF evolution, and actual irrigation timing and water depth applied; this adjustment is seldom found in most irrigation scheduling models available in the market. We have assumed that ETp data retrieved from weather stations and actual irrigation water application are trustworthy, since modern irrigation equipment provides automatic digital operation registering options (date, time, water volume applied on each field irrigated section), including data transmission by radio frequency or through the Internet, thus reducing human intervention on data handling.

3. Irrigation system maintenance, improvements, and spare parts replacement

Routine irrigation equipment maintenance protocols are needed for sustainable achievement of the potential economic return of investments, by ensuring the timely, complete, uniform, and efficient water supply to the crop. In most field systems, regardless its size or irrigated crop value, maintenance protocols are seldom implemented in full, and generally are only addressed when major system failures are detected, affecting crop yield and fruit quality, due to water supply interruptions during the repairing time span [39]. Maintenance is an important, though often overlooked, operation to extend not only the trouble-free life of the system itself but to maximize returns on investment. Preventative rather than corrective maintenance is more economical and less traumatic. The implementation of a maintenance program for drip irrigation systems will keep the system operating at peak performance and increase the system's work life expectancy. The best way to determine if the maintenance program implemented is effective is to constantly monitor and record the flow rate and pressures in the system [3].

3.1 Winter maintenance protocol

It is one of the most important maintenance activities, to be performed during the postharvest winter period; if the total winter rainfall is below the average value of the area, it is necessary to operate the irrigation system at the beginning of spring, before crop budbreak (permanent orchards) or emergence (annual crops), when the spring root activity is initiated, in order to start the irrigation season with a soil water depth equivalent to its field capacity.

Pumps, filters, and valves are dismantled in the control room, as well as the entire electrical installation, including power boards, irrigation programming boards, filter back-washing boards, and all the fertilizer preparation and injection systems. In the field, mains, submains, manifolds, and irrigation laterals are washed to evacuate any sediment that may have precipitated and the emitters are revised to change those that are in poor condition. In persistent fruit orchards species, it may be necessary to continue watering in the winter, so this maintenance operation is performed after a significant rain.

The goal of the winter maintenance is to ensure that at the start of the new irrigation season, all equipment components are in optimal operative condition. The cost of this winter maintenance operation, including the cost of some spare parts that need to be replaced, plus the replacement cost of filtering media (quartz sand, meshes, and filter disks), generally represents 2–3% of the original irrigation equipment investment. Once the irrigation equipment has been reassembled after this winter maintenance, it is necessary to calibrate its operation, in terms of the emitters' discharge uniformity, operating pressures across the whole hydraulic network, and elimination of water leaks [40].

3.2 Irrigation equipment routine operation maintenance

Throughout the irrigation season, implementing a daily maintenance protocol for the irrigation equipment components is required, basically consisting of the analysis of the registered operation information provided by volume totalizers, flow measurements, and operation times for each irrigated sector, for early detection of eventual anomalies in its operation.

The goal is to always keep within a range of variation that does not exceed 5% of the pressure and discharge values established in the original design of the equipment. If any of these two parameters deviate above or below this range, at any point of the irrigation network, it is necessary to find the failure point or section and repair it immediately, to maintain the correct supply of water to the crop.

The most frequent problems to find during the irrigation season are partial emitter clogging, irrigation hydraulic valve elasticity reductions, leading to incomplete opening or closing, breakages and leaks in the water distribution pipes, filters' inadequate cleaning, malfunctioning of electrically operated pilot valves, pump efficiency reductions, and mechanical damage of laterals due to field operations (labor, animal, or machinery) or rodent damage.

3.3 Periodic maintenance

At least once every 2 weeks, the following maintenance procedure is mandatory:

1. Flush all the laterals by opening end plug 1–5 in a series; then close them 1–5 in the same sequence allowing flushing for 3 min until clean water starts flowing.

2. Flush each submain at the end of every section (shift) till dirt-free clear water starts flowing.

3. Check inlet and outlet filter pressures. Remove slurry from sand filtration media with back flush at every 5 h; flush screen/disc filter.

4. Take out the element of screen/disc filter and clean it thoroughly. Open the lid of sand (media) filter, allow the water to come out through it, for thoroughly separating accumulated foreign material with media (sand) for recharging its filtering capacity.

In most situations, irrigation equipment malfunctions develop dynamically, leading to increasing expenditures for its solutions; thus, early detection of operational issues not only prevents negative impacts on crop yield and quality, but also in repairing costs. Comparing data on emitter discharge, end of lateral pressure, pressure at main valves, and water flow after filters, registered at least weekly, has proven to be an excellent method for early detection of irrigation equipment deficiencies.

Many irrigation systems are provided with operational registering options; however, the systematic analysis of this information is seldom included in the routine activities of field decision-making personnel. Available technologies enable the use of automatic cellphone alarms, triggered when the system operation deviates from specific preset discharge or pressure parameters. Data on actual irrigation system performance are seldom considered as a valuable crop production input; thus, a major educational effort is due to fully make use of these system capabilities, at a very low cost and in just a few training hours.

4. Experiences

We have selected data from just one field using our irrigation scheduling platform as an example, to fully present its many applications. The authors have implemented irrigation scheduling professional consulting since 1982 in more than 150 horticultural plantations in Chile, Mexico, Peru, and Argentina, on a cultivated area estimated at 5500 hectares. Concepts, parameters, coefficients, and computer programs, developed and published extensively in scientific and professional journals and presented at countless congresses, courses, and workshops, today serve as the basis for the correct use of irrigation systems in horticultural plantations, carried out also by many other professionals and technicians in different countries; selected irrigation scheduling publications by the senior author are available on Internet [41].

PLOT						
Cultivar	**C. sauvignon**	**Merlot**	**Syrah**	**Sauvignon blanc**	**Chardonay**	**C. sauvignon**
Weather station	1	2	3	4	5	6
Area (has)	9.35	6.68	11.24	10.26	14.23	9.45
Soil type	1	2	3	3	1	2
% Sand	60.20	26.8	18.5	18.5	60.2	26.8
% Loam	34.40	49.2	41.3	41.3	34.40	49.2
% Clay	5.40	24.1	40.2	40.2	5.40	24.1
Field Capacity (Vol. %)	19.84	29.77	38.21	38.21	19.84	29.77
Irrigation sistem number	1	1	1	2	2	2
Dripper flow (L/Hr)	1.2	2.0	2.0	2.0	1.2	2.0
Laterals/Line	2	1	1	1	2	1
Drippers/Ha	14815	4000	4000	4000	14815	4000
Discharge (m3/Hr)	17.78	8.00	8.00	8.00	17.78	8.00
Phenology:						
Budbreak	18-09-2018	14-09-2018	31-08-2018	15-08-2018	18-08-2018	18-09-2018
Flowering	14-10-2018	08-10-2018	04-10-2018	30-09-2018	27-09-2018	18-10-2018
Midseason	04-11-2018	31-10-2018	29-10-2018	14-10-2018	08-10-2018	06-11-2018
Veraison	08-01-2019	29-12-2018	26-12-2018	26-12-2018	18-12-2018	08-01-2019
Maturity	03-03-2019	06-03-2019	10-03-2019	31-01-2019	18-02-2019	10-03-2019
Last Irrigation	19-04-2019	14-04-2019	14-04-2019	14-04-2019	20-04-2019	21-04-2019
Kc:						
Kc min	0.15	0.16	0.17	0.18	0.19	0.20
Kc max	0.71	0.69	0.65	0.75	0.70	0.71
Kc end of harvest	0.60	0.58	0.60	0.70	0.68	0.60
Kc end of season	0.10	0.10	0.11	0.12	0.13	0.14
MAD* (%)	5.0	10.0	15.0	15.0	5.0	10.0

Table 1.
Vineyard data relevant for irrigation scheduling.

Data for a 61.31-hectare vineyard, with two independent drip irrigation systems, three sectors each, planted in three different soils, with five different cultivars and two climate evaporative demand conditions are presented (**Table 1**).

Figures 3 and **4** present data on irrigation for Cabernet sauvignon for plots 1 and 6 (different evaporative demand conditions, due to site topographic positions within the vineyard, as well as different soil hydrodynamic characteristics), for a time frame from November 15 to December 31, 2018; the platform enables the user to select data for any plot or plot combination, for any time span, from 1 day to the whole irrigation season. Daily comparisons between calculated ETp (red columns) and actual water depth applied (blue columns) are provided in graphic format, indicating a correct operation of the irrigation system throughout both dates, with the exception of December 25th, when no irrigation was performed, followed by two intensive irrigation days, to recover the difference. Water depths applied during the specified time span are also provided, comparing calculated ETp and water depth applied. This information is a helpful tool to decision-making for water recovery or withhold, aimed to keep a constant soil water content in the root zone.

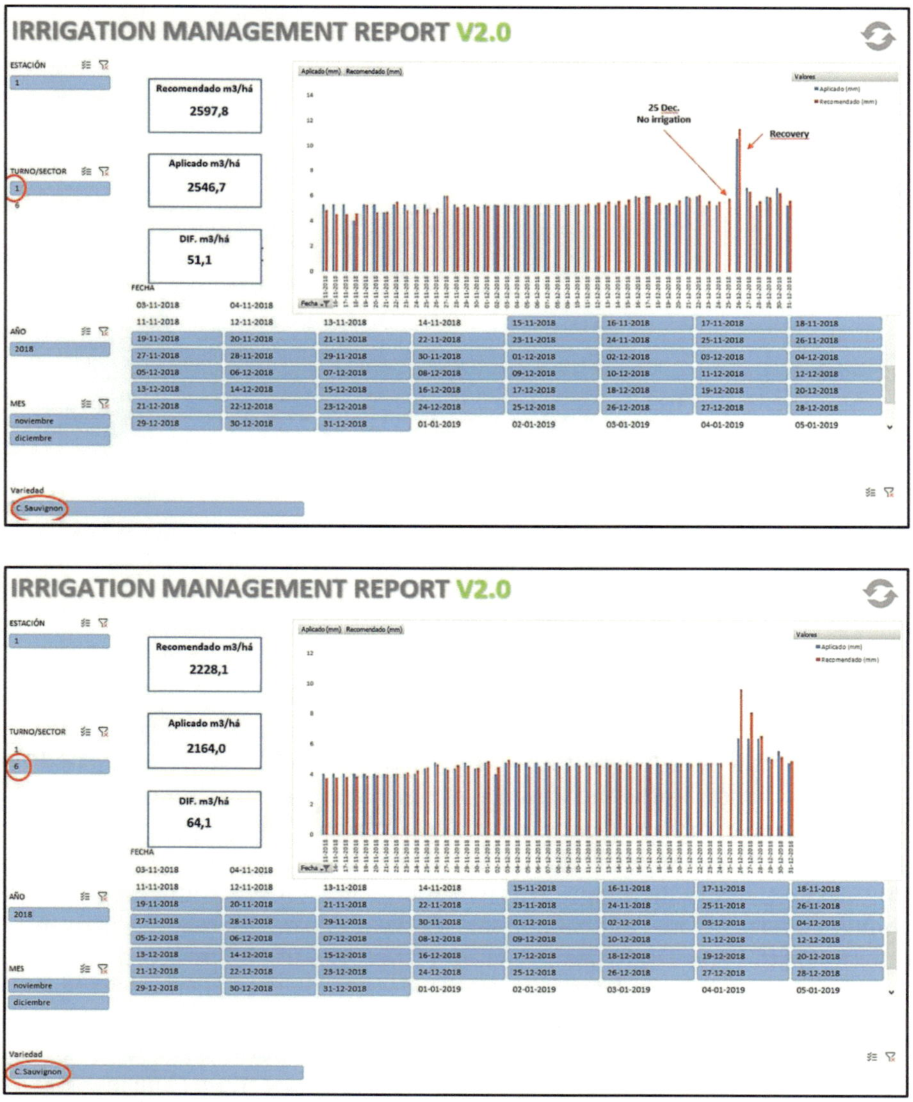

Figure 3.
Irrigation scheduling for two Cabernet sauvignon plots, November 15–December 31, 2018.

Figure 4.
Irrigation scheduling for Chardonnay and Merlot plots, November 15–December 31, 2018.

A 15.1% difference on ETp between plots 1 and 6 accounts for climatic evaporative water differences between both plots. Three years before the implementation of the irrigation scheduling platform at this vineyard, both plots were irrigated with identical water depths and timings; as a result, significant differences in grape yield, average berry size, and wine organoleptic characteristics were obtained. These differences are almost nil for the last two vintages. Also, annual water and energy savings, due to the adoption of irrigation scheduling, account for 34.7%.

Similar data for plots 2 and 5 (cultivars Merlot and Chardonnay, respectively) are presented in **Figure 4**. Irrigation scheduling procedures follow a consistent concordance for daily calculated ET and actual water depth applied.

Between these dates, differences in water depth applied, as compared to calculated ETp, are 3.79% for Merlot and 0.21% for Chardonnay; however, the difference between both cultivars accounts for 340 m^3/hectare. Considering the whole season (data not shown), the calculated ETp difference between both cultivars is 1.216 m^3/hectare, equivalent to 19.7%, due to differences in the onset of each phenology dates, which significantly modify its respective Kc = f(t) functions. An irrigation scheduling strategy adequate for Merlot, applied into Chardonnay, will result in overirrigation, excessive canopy vigor, and poor wine organoleptic characteristics, as well as unnecessary water end energy costs in this last plot.

At the same vineyard, a different situation regarding irrigation scheduling was detected; **Figure 5** presents data for plots 3 and 4 (cultivars Syrah and Sauvignon blanc, respectively). Irrigation scheduling between December 1, 2018 and January 18, 2019 consisted on a daily 8 h unique irrigation event, regardless of actual ETp, with no irrigation taking place in December 25.

For the Syrah plot, on the average between these dates, no differences between calculated ETp and actually applied water depth are detected, but if each day is considered separately, overirrigation took place during 20 days, and during 19 days, the plot was underirrigated. This time span includes the berry veraison

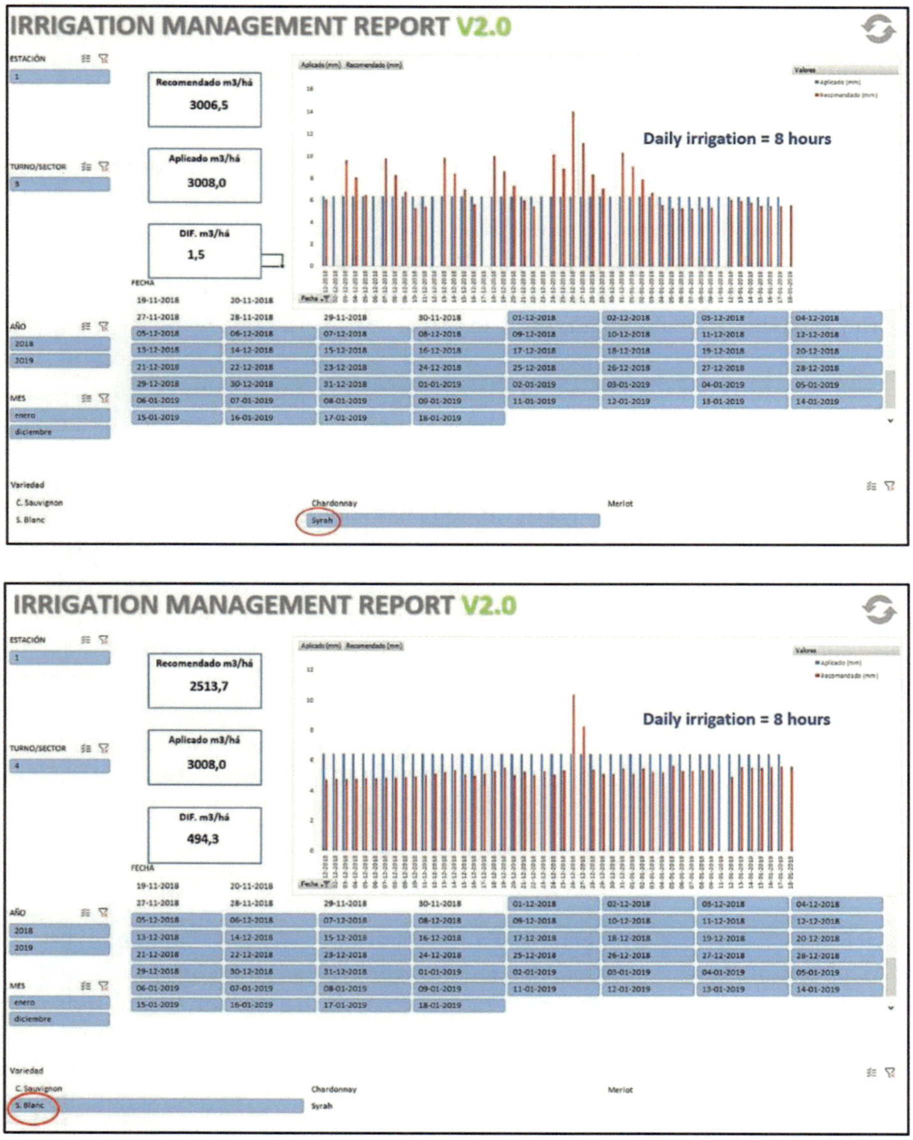

Figure 5.
Irrigation scheduling for Syrah and Sauvignon blanc plots, December 1, 2018–January 18, 2019.

to maturity stage; this irrigation strategy produced 14.6% larger than targeted average berry size (data not shown) and possibly, with expected negative effects on wine organoleptic indicators. In the Sauvignon blanc plot, overirrigation took place throughout these dates, except for December 25 and 26; for this cultivar, the phenology stage corresponds to berry final maturity, which was delayed by 10 days (harvesting date was January 31'); also, 18–20% of the berries cracked due to excessive irrigation and *Botrytis cinerea* affected a significant number of grape clusters.

5. Constraints in the adoption of smart water management tools and strategies into farming operations

There is a generalized feeling among farmers and field extensionists in relation to irrigation, who almost unanimously and systematically consider today that this agronomic practice is the major limiting factor in crop productivity in most farms. Efforts done to effectively improve water productivity are affected by two main constraints in the adoption of smart water management applications in field-pressurized irrigation systems:

1. The relatively low cost of water/energy, in relation to other production inputs, which determine a negative stimulus to actively implement irrigation scheduling and equipment maintenance, and

2. Inadequate knowledge on the actual relation between crop yield/quality and the correct water supply strategy, in terms of the effective water depth applied on each irrigation event, the importance of correct water application timing and the impact of uneven water distribution over the irrigated field, due to improper equipment maintenance.

The first constraint is being painfully addressed as a result of decreasing irrigation water availability, due to climatic global change, but the second constraint requires an urgent upgrade in irrigation decision-makers' knowledge and professional abilities, at the farm level. Highly motivated extensionists, with specific quantitative goals to address these constraints are needed in most agricultural areas, in the scope of well-financed collective policy schemes, to obtain the highest economic return for each water drip available. Efforts to provide short-term, accredited, and practical courses on irrigation system performance and maintenance at the farm level to operators and agronomists are urgently needed in most agricultural irrigated areas.

6. Conclusions

Integration of irrigation scheduling and irrigation system maintenance concepts and techniques is a most needed technology to be adopted by agricultural stakeholders, providing data and orientations to optimize the benefits of irrigation investments, influencing both crop yield and quality, as well as by significant reductions in water, energy, and repairing costs. Professional specialized advice on the operation of the irrigation equipment, including daily irrigation scheduling, irrigation equipment maintenance and training, and control of irrigation system and field personnel operational performance throughout the season is highly recommended.

Implementation costs of continuous irrigation scheduling services and system maintenance protocol analysis, including field personnel training, are almost irrelevant (in the range of U$ 20–U$ 90/hectare-year), in the scope of crop annual production costs. Moreover, incorporation of new Internet of Things applications for sensor collection of field data and its real-time analysis with increasingly powerful graphic software, using big data analysis tools, indicates that further cost reductions and increasing applications of irrigation scheduling to farming can be expected.

Author details

Luis A. Gurovich[1]* and Luis Fernando Riveros[2]

1 Agriculture and Forestry School, Universidad Catolica de Chile and G&A Consultants S. A., Santiago, Chile

2 Facultad de Medicina Veterinaria y Agronomía, Universidad de Las Americas and G&A Consultants S. A., Santiago, Chile

*Address all correspondence to: lgurovich@gya.cl

References

[1] Hamdy A. Water use efficiency in irrigated agriculture: An analytical review. In: Lamaddalena N, Shatanawi M, Todorovic M, Bogliotti C, Albrizio R, editors. Water Use Efficiency and Water Productivity: WASAMED Project. Bari: CIHEAM; 2007. pp. 9-19 (Options Méditerranéennes: Série B. Etudes et Recherches; n. 57). 4. WASAMED (WAter SAving in MEDiterranean agriculture) Workshop, 2005/09/30-2005/10/04, Amman (Jordan). Available from: http://om.ciheam.org/om/pdf/b57/00800773.pdf

[2] Sustainable Agriculture Initiative Platform. Water Conservation Technical Briefs TB 6—Irrigation Scheduling. 2010. Available from: http://www.saiplatform.org/uploads/Modules/Library/SAI%20Technical%20Brief%206%20%20%20Irrigation%20Scheduling.pdf [Accessed: January 9, 2019]

[3] Gurovich LA, Riveros LF. Estudio básico "Diagnóstico de la Eficiencia de Aplicación del Riego en Chile". Publication CNR-0430 V2. 2014. 308 p. Available from: http://bibliotecadigital.ciren.cl/bitstream/handle/123456789/26337/CNR-0430-V2.pdf?sequence=2&isAllowed=y

[4] Goyal MR, Panigrahi B, Panda SN. Micro Irrigation Scheduling and Practices. Series: Innovations and Challenges in Micro Irrigation. USA: Apple Academic Press; 2017. 384 p

[5] Fernández JE. Plant-based methods for irrigation scheduling of woody crops. Horticulturae. 2017;**3**:35-72

[6] Irrigation Crop Diversification Corporation. Irrigation Scheduling Manual. 2017. Available from: http://irrigationsaskatchewan.com/icdc/wp-content/uploads/2017/05/Irrigation-Scheduling-manual-June2017.pdf [Accessed: January 10, 2019]

[7] Pereira LS. Chapter 1: Inter-relationships between irrigation scheduling methods and on-farm irrigation systems. In: Irrigation Scheduling: From Theory to Practice–Proceedings of the ICID/FAO Workshop on Irrigation Scheduling. Water Report Series No. 8. 1996. Available from: http://www.fao.org/docrep/w4367e/w4367e0c.htm

[8] Kisekka I, Migliaccio KW, Dukes MD, Schaffer B, Crane JH. Evapotranspiration-Based Irrigation Scheduling for Agriculture. AE457. Gainesville: University of Florida Institute of Food and Agricultural Sciences; 2009. Available from: http://edis.ifas.ufl.edu/ae457

[9] Allen RG, Pereira LS, Raes D, Smith M. Crop evapotranspiration: Guidelines for computing crop water requirements. United Nations FAO, Irrigation and Drainage, N.Y., Paper No. 56. Rome: Food and Agriculture Organization of the United Nations; 1998. Available from: http://www.fao.org/docrep/X0490E/ x0490e00.htm. ISBN: 92-5-104219-5

[10] Cai J, Liu Y, Lei T, Pereira LS. Estimating reference evapotranspiration with the FAO Penman-Monteith equation using daily weather forecast messages. Agricultural and Forest Meteorology. 2007;**145**:22-35

[11] Kamienski CA, Soininen JP, Cinotti TS. SWAMP: An IoT-based smart water management platform for precision irrigation in agriculture. In: IEEE Conference on Global IoT Summit 2018 (GIoTS'18); 2017. pp. 1-6. Available from: https://ieeexplore.ieee.org/abstract/document/8534541

[12] Singh KG, Kaur P. Chapter 7: Evapotranspiration estimations using climatological approaches. In: Singh KG, Goyal MR, Rudra RP, editors.

Best Management Practices for Drip Irrigated Crops. Research Advances in Sustainable Micro Irrigation Series. Vol. 6. USA: Apple Academic Press Inc; ISBN: 13:978-1-4987-1482-2

[13] Zeleke KT, Wade LJ. Evapotranspiration estimation using soil water balance, weather and crop data. In: Labetski L, editor. Evapotranspiration—From Measurements to Agricultural and Environmental Applications. 2012. ISBN: 9533075120

[14] Hobbins M, Huntington JL. Chapter 42: Evapotranspiration and evaporative demand. In: Singh VP, editor. Handbook of Applied Hydrology. 2nd ed. USA: McGraw-Hill Education; 2016. ISBN: 13:978-0071835091

[15] Dec D, Dorner J. Spatial variability of the hydraulic properties of a drip irrigated andisol under blueberries. Journal of Soil Science and Plant Nutrition. 2014;**14**:589-601

[16] Joy L, Starks JL, Philip J, Pratt PJ, Mary Z, Last MZ. Concepts of Database Management. 9th ed. USA: Cengage Learning Print; 2018. ISBN: 9781337093422

[17] Casadesús J, Mata M, Marsal J, Girona J. A general algorithm for automated scheduling of drip irrigation in tree crops. Computers and Electronics in Agriculture. 2012;**83**:11-20

[18] Queensland Department of Agriculture. 2019. Irrigation Scheduling and Crop Water Use Estimation Web site. Available from: https://waterschedpro.net.au/

[19] Snyder RL, Geng S, Orang M, Sarreshteha S. Calculation and simulation of evapotranspiration of applied water. Journal of Integrative Agriculture. 2012;**11**:489-501

[20] Zhe G, Zhiming Q, Liwang M, Dongwei G, Junzeng X, Quanxiao F, et al. Development of an irrigation scheduling software based on model predicted crop water stress. Computers and Electronics in Agriculture. 2017;**143**:208-221

[21] Saxton KE, Rawls WJ, Romberger JS, Papendick RI. Estimating generalized soil-water characteristics from texture. Soil Science Society of America Journal. 1986, 1986;**50**:1031-1036. Available from: https://dl.sciencesocieties.org/publications/sssaj/abstracts/50/4/SS0500041031 http://resources.hwb.wales.gov.uk/VTC/env-sci/module2/soils/soilwatr.htm

[22] Zur B. Wetted soil volume as a design objective in trickle irrigation. Irrigation Science. 1996;**16**:101-105

[23] Simonne E, Gazula A, Hochmuth R, De Valerio J. Chapter 7: Water movement in drip irrigates sandy soils. In: Goyal RM, editor. Sustainable Practices in Surface and Subsurface Micro Irrigation. Vol 2. USA: CRC Press; pp. 183-210. ISBN:13: 978-1-77188-017-6

[24] Zotarelli L, Dukes MD, Romero CC, Migliaccio KW, Morgan KT. Step by step calculation of the Penman-Monteith Evapotranspiration (FAO-56 method). University of Florida Extension Paper AES 459. 2013. Available from: http://edis.ifas.ufl.edu/ae459

[25] Zermeño-González AA, Melendres-Alvarez AI, Fuerte-Mosqueda LA, Munguia-López JP, Ibarra-Jiménez L. Evapotranspiration rate of a vineyard and its relation to the reference of the FAO Penman-Monteith method. Agrociencia. 2017;**51**:1-12

[26] Villagra P, García de Cortázar V, Ferreyra R, Aspillaga C, Zúñiga C, Ortega-Farias S, et al. Estimation of water requirements and Kc values of 'Thompson Seedless' table grapes grown in the overhead trellis system, using the Eddy covariance method. Chilean Journal of Agricultural Research. 2014;**74**:213-218

[27] Allen RG, Pereira LS, Smith M, Raes D, Wright JL. FAO-56 dual crop coefficient method for estimating evaporation from soil and application extensions. Journal of Irrigation and Drainage Engineering. 2005;**131**:1-17

[28] Kamble B, Irmak A, Hubbard K. Estimating crop coefficients using remote sensing-based vegetation index. Remote Sensing. 2013;**5**:1588-1602

[29] Marin FR, Angelocci R, Nassif DS, Costa LG, Vianna MS, Carvalho KS. Crop coefficient changes with reference evapotranspiration for highly canopy-atmosphere coupled crops. Agricultural Water Management. 2016;**163**:139-145

[30] Smith M. Climwat for Cropwat: A Climatic Database for Irrigation Planning and Management FAO, Rome (Italy). FAO, Rome (Italy): Land and Water Development Div., Research and Technology Development Div.; 1993. Available from: http://www.fao.org/land-water/databases-and-software/climwat-for-cropwat/en/

[31] Kopp K, Hoover J. Irrigation System Maintenance. All Current Publications Utah State University. Paper 824. 2004. Available from: https://digitalcommons.usu.edu/extension_curall/824 [Accessed: January 10, 2019]

[32] Prabhugouda P, Biradar P, Bhagawathi AU, Hejjegar IS. Review on leaf area index of horticulture crops and its importance. International Journal of Current Microbiology and Applied Science. 2018;**7**:505-551

[33] Ghandour A, Snyder RL, Frame K, Eching S, Temesgen B, Davidoff B. Converting Kc values between ETo and ETr. In: World Environmental and Water Resource Congress 2006. Available from: https://ascelibrary.org/doi/pdf/10.1061/40856%28200%29258 [Accessed: January 9, 2019]. ISBN: 9781771885522

[34] Henggeler JC, Guinan CP, Travlos J. Procedure to easily Fine-Tune Crop Coefficients for Irrigation Scheduling. 2008. Available from: https://www.irrigation.org/IA/FileUploads/IA/Resources/TechnicalPapers/2008/ProceduresForEasilyFine-TuningCropCoefficientsForIrrigationScheduling.pdf [Accessed: January 10, 2019]

[35] Poblete-Echeverría C, Sepúlveda-Reyes D, Zúñiga M, Ortega-Farías S. Grapevine crop coefficient (Kc) determined by surface renewal method at different phenological periods. Acta Horticulturae. 2017;**1150**:61-66

[36] Schwartz MD, editor. Phenology: An Integrative Environmental Science. 2nd ed. Switzerland: Springer; 2003. 610 p. ISBN: 13:978-9400769243

[37] Steven T, Hartley S, Burns KC. Global patterns in fruiting seasons. Global Ecology and Biogeography. 2008;**17**:648-657

[38] Williams LE, Ayars JE. Grapevine water use and the crop coefficient are linear functions of the shaded area measured beneath the canopy. Agricultural and Forest Meteorology. 2005;**132**:201-211

[39] Netafim. Guidelines for Irrigation Systems Maintenance. 2011. Available from: https://www.netafimusa.com/wp-content/uploads/2016/08/Preventive-Maintenance-Guide.pdf

[40] Mostafa H, Thörmann HH. On-farm evaluation of low-pressure drip irrigation system for smallholders. Soil and Water Research. 2013;**8**:87-95

[41] Gurovich L. Selected papers on irrigation by the senior author. Available from: https://scholar.google.cl/scholar?q=Gurovich+Irrigation+Scheduling&hl=es&as_sdt=0&as_vis=1&oi=scholart

Chapter 3

Integrating Remote Sensing Data into Fuzzy Control System for Variable Rate Irrigation Estimates

Willians Ribeiro Mendes, Fábio Meneghetti U. Araújo and Salah Er-Raki

Abstract

Variable rate irrigation (VRI) is the capacity to vary the depth of water application in a field spatially. Developing precise management zones is necessary to efficient variable rate irrigation technologies. Intelligent fuzzy inference system based on precision irrigation knowledge, i.e., a system capable of creating prescriptive maps to control the rotation speed of the central pivot. Based on the VRI-prescribed map created by the intelligent system of decision-making, the pivot can increase or decrease its speed, reaching the desired depth of application in a certain irrigation zone. Therefore, this strategy of speed control is more realistic compared to traditional methods. Results indicate that data from the edaphoclimatic variables, when well fitted to the fuzzy logic, can solve uncertainties and non-linearities of an irrigation system and establish a control model for high-precision irrigation. Because remote sensing provides quick measurements and easy access to crop information for large irrigation areas, images will be used as inputs. The developed fuzzy system for pivot control is original and innovative. Furthermore, the artificial intelligent systems can be applied widely in agricultural areas, so the results were favorable to the continuity of studies on precision irrigation and application of the fuzzy logic in precision agriculture.

Keywords: fuzzy control, variable rate irrigation, center pivot control, remote sensing, decision support system

1. Introduction

Availability of water is one of the basic conditions for life on planet Earth. However, it is a limited resource, currently at risk of extinction. Global population growth, climate change and demand from several economic sectors such as industry and agriculture put into question the availability of drinking water to all living beings on planet. In particular, irrigated farming is one of the sectors that consume more water per day, and can reach 90,000 liters/hectare, while the average consumption per capita in Brazil is 162 liters per day [1].

The United Nations Food and Agriculture Organization (FAO) predicts that global demand for food by the year 2050 will increase by at least 60% above 2006 levels, and in order to meet this demand it would need to double or triple agricultural production. However, most of the food production increase must have come from yield increases [2]. According to [3], the adoption of irrigated agriculture makes it possible to increase productivity and diversify agricultural crops. However, there is a limitation in water resources and, therefore, the use of water in agriculture needs to be more efficient.

The [4–6] present an overview of precision agriculture. The authors state that the term can be used in everything that refers to activities performed more accurately by means of electronic systems; however, they make a note regarding the applications of inputs uniformly, which would be only conventional systems and not deal with the spatial variability of crops. Automation and instrumentation solutions are required for better application of inputs, and in order to achieve a distinct water management in each sector of a planted area, irrigation systems must perform water application taking into account the spatial variability of the crop and the soil so that the maximum efficiency of the crop can be reached [7].

Authors such as [8–13] discuss some solutions for water application using spatial correction and conclude that the central or linear pivot or irrigator are particularly suited to the precision irrigation condition, especially because of their current levels of automation and large area reached by the pivot. However, the major limitation for the adoption of irrigation that complies with spatial and time variability, usually called variable rate irrigation, is associated with the development of great irrigation management.

The availability of sensors is currently a constraint to the automation of irrigation control, and it is expected that the requirements of advanced process control for irrigation also fosters the development of new sensors. In [14] brings a review of the existing literature on advanced process control in irrigation and its requirements of sensors and adaptability to the field conditions, besides discussing the obstacles in area sensing.

In order to deliver detailed spatial and temporal information regarding soil and crop response to varied management practices and dynamic environmental conditions, and to avoid the time-consuming process for installing and maintaining sensors over each field, the use of remote sensing techniques has been improving in precision agriculture [15, 16]. According to the authors, remote sensing images are already widely used and proved to do a good prediction on required irrigation amount for each type of crops. Remote sensing by satellite has been very promising in on-field monitoring, but still presents problems such as accuracy, cloud coverage, and the high cost to obtain good spatial resolution [17].

The application of process control techniques for variable rate irrigation has recently been reviewed in [18–24]. Artificial intelligence (AI) can be applied in an interdisciplinary way, besides bringing about a paradigm shift of how we understand agriculture today. Solutions in AI technology not only enable farmers to do more with less, but also improve quality and ensure a faster introduction into the market.

AI technologies assist farmers in soil analysis and crop health, among others, besides saving time and allowing them to grow the right crop at each season, thereby maximizing the crop production. In this context, tools with knowledge representation and reasoning about imprecision present as a feasible alternative. In this way, fuzzy logic allows intelligent computational systems to "reason", considering aspects inherent to uncertainty and realistic processes. Moreover, it is a very interesting methodology to be applied in decision making, because it is possible to model perceptions and preferences similar to the style of a human being.

Decision support systems are tools that can be used in fuzzy set theory [25] to provide a conceptual framework for representing knowledge and reasoning about imprecision and consequent uncertainty. The fuzzy set provides adequate tools for modeling and dealing with expert rules [26]. By modeling linguistic variables in the form of fuzzy sets, it was possible to transform expert rules into mathematical terms, and in addition, fuzzy set theory offers a wide variety of operators that can aggregate and combine these rules. The application of linguistic variables and fuzzy conjunction methods provide an adequate method to model the human reflection process and, in so doing, make the interface of these systems simpler and more natural as planning tool on the farm by the manager or farmer.

The fuzzy decision support system is considered useful due to its interactive nature, flexibility in approach and evolution of the graphical characteristics, and can be adopted for any similar situation to classify the alternatives. More often, ambiguity in agricultural decision-making is aggravated by inaccuracy and intuition. The ability of fuzzy systems to deal with complex systems can help farmers to make better decisions in agricultural processes [27]. There is a very significant advantage in using fuzzy decision-making systems for the variable rate irrigation process: the advantage of not needing the full amount of relevant information by simply selecting the variables that play the role of the irrigation calculation according to [28].

This chapter is organized as follows, the present section aims to contribute to a refinement of the studies on the application of fuzzy control systems for the exploration of precision irrigation modeling and management. In the next section, we provide a literature review of the latest related research, divided into three subsections, namely: (a) the most important concepts for understanding the main characteristics in a central pivot irrigation system; (b) concepts fundamental to the understanding of fuzzy logic, relevant to the structuring and development of the intelligent irrigation system and (c) remote sensing. In Section 3, we thoroughly describe the basic mathematical framework that involves the three techniques. Finally, a representative case study on the intelligent control of variable rate irrigation systems is presented.

2. Precision agriculture

Available bibliographies give different names to describe the concept of precision agriculture such as spatially prescriptive agriculture, computer farming, satellite farming, high technology for sustainable agriculture, soil specific crop management or site-specific crop management. It is considered a revolutionary approach to improving resource management and sustainable agricultural development and is a promising technology. [29–33].

Precision agriculture studies were started in countries such as the USA, Canada, Australia, and Germany, besides the Western Europe, in the mid-1980s, and only began to receive great interest as a new experimental tool in the 1990s [29]. In [34] is define the specific management of a study zone as the electronic monitoring and control applied to data collection, information processing and decision support for the temporal and spatial allocation of inputs for agricultural production. The specific control zone, as shown in **Figure 1**, is spatially defined by soil elements, crop type, pests and other elements required for efficient management of inputs.

Technologies on agricultural production are expected to impact in two areas: profitability for producers and ecological/environmental for the public. Increased costs with water, fertilizer and pesticides, coupled with environmental concerns, lead to a growing acceptance of the concept of specific management of an operating zone.

Figure 1.
Different management zones within the same planting area. Source: Embrapa. (https://www.macroprograma1.cnptia.embrapa.br/redeap2).

2.1 Variable rate irrigation (VRI)

Variable rate irrigation (VRI) is a specific management tool used to apply the adequate amount of water in the sectors or zones of a planting area, for example **Figure 2**, presents control regions where the zones in reddish colors need more water, and those of bluish colors are with the humidity within the limits that the plant needs. The development of the prescription of variable rate irrigation is a field of active research, studied in [13, 14, 35].

Once established, prescriptions can, within a management variability, remain fixed, or these zones can dynamically change a small number of times during a growing season. Characteristics of crops and soil type are the main factors that contribute to determine the space and time variability of a planted area. This information is incorporated into a geographic information systems (GIS) database and, therefore, used for interpretation and decision support [36].

2.1.1 Irrigation system

Irrigation systems are a set of techniques aimed to distribute water to crops in adequate quantities in order to promote appropriate plant development with a

Figure 2.
Spatial variability of irrigation water needs. Source: VALLEY. (http://ww2.valleyirrigation.com/valley-irrigation/pt/tecnologia-de-comando/rega-de-taxa-vari%C3%A1vel/vri-controle-de-velocidade).

minimum of water consumption [37]. Irrigation systems can be divided into two subsystems: catchment and application subsystem. The way the water is applied depends on different methods of application, and each has its specificities. They are divided into three groups: surface irrigation, localized irrigation, and sprinkler irrigation.

- Surface irrigation: the water from the distribution system (channels and pipes) to any point of infiltration within the area to be irrigated is made directly on the surface of the soil. They are classified as infiltration furrows and flooding or submersion;

- Localized irrigation: the water is applied directly on the root area, with small intensity and high frequency. Classified as: micro-sprinkler and drip;

Sprinkler irrigation is the method of irrigation in which water is sprayed on the surface of the land, like a rainfall, because the water jet is fractioned in drops. They are classified as: conventional spraying, central pivot, self-propelled, and linear system. Since this is the scope of this chapter, central pivot irrigation will be further detailed.

2.1.2 Central pivot

Among the sprinkler systems, the central pivot has been used with relative success due to the lower labor demand [37, 38]. It was first built in 1948 by Frank L. Zybach, who sent the invention for analysis, finally patented in 1952 in Colorado, United States (see **Figure 3**). In 1954, Zybach sold the manufacturing rights to the American company Valley, located in the State of Nebraska. In 1968, the Lindsay Company also started to produce pivots, and currently both companies share the leadership of the world market of pivots.

The speed of the lateral displacement of a central pivot is controlled in the last tower, which is established by a timer, installed in the central control box of the pivot, which controls the time of activation and the stop of the motor of the last tower. For example, the condition in which the motor standstill time is equal to the movement time corresponds to the setting of 50% of the maximum speed set by the timer control percentage. At maximum speed of 100%, the motor of the last tower is continuously moving [37, 39].

Irrigated agriculture does not allow reductions in crop productivity due to lack or excess of applied water. The application of little water (deficit irrigation) can be an obvious waste, since production could not obtain the expected benefit. On the other hand, the excessive application is much more destructive, because soil saturation occurs, which prevents its aeration and leaches the nutrients, inducing a higher rate of evaporation and salinization [40]. So, it is important to develop an irrigation scheduling program for deciding when and how much to irrigate. For this purpose, we used the fuzzy logic system to simulate the amount and the frequency of irrigation needed.

2.2 Fuzzy logic

Fuzzy sets theory was introduced in 1965 by the Iranian mathematician Lotfi Asker Zadeh, a professor at the University of Berkley, USA [41], especially intended to offer a mathematical treatment to some subjective linguistic terms such as "approximately" and "around", among others. This would be a first step in programming and storing vague concepts in computers, making it possible to produce calculations with inaccurate information, such as the human being [42].

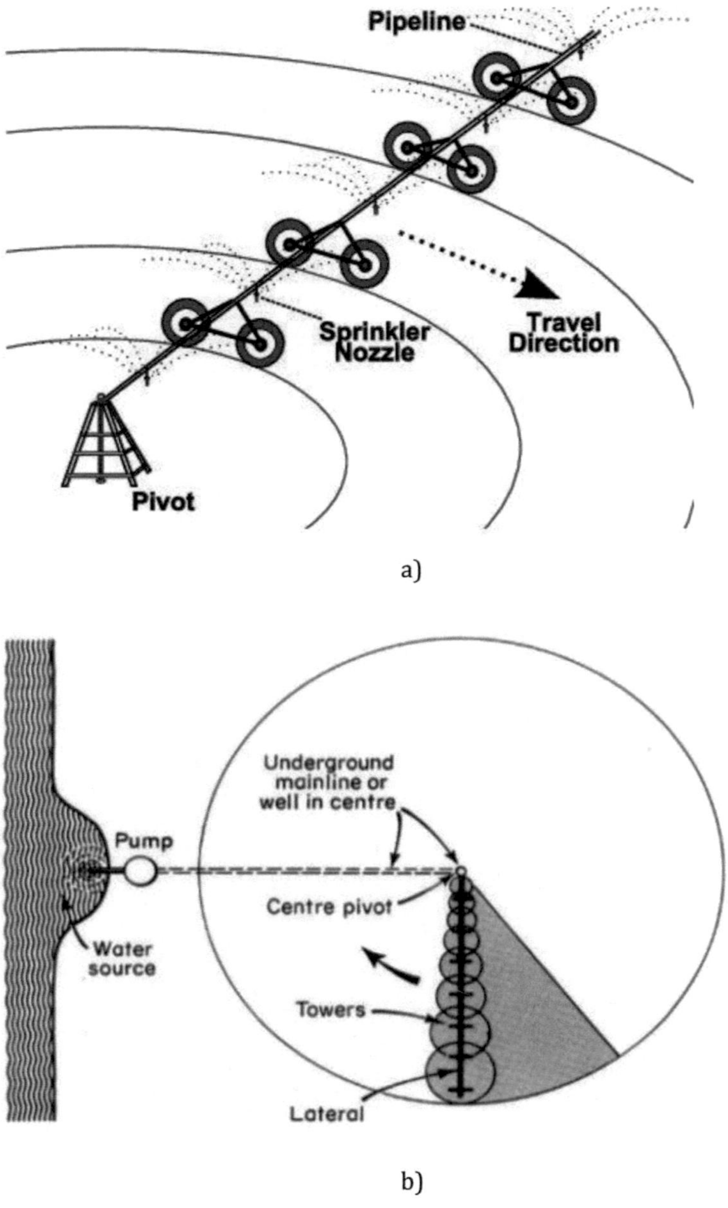

Figure 3.
Structure of a central pivot. (a) Basic components, and (b) irrigated land. Source: Adapted from [38].

In other words, while decision making in classical theory would be like Eq. (1), fuzzy logic would be like Eq. (2) [43].

$$f(x) = \begin{cases} 1\ if,\ and\ only\ if,\ x \in A \\ 0\ if,\ and\ only\ if,\ x \notin A \end{cases} \tag{1}$$

$$f(x) = \begin{cases} 1\ if,\ and\ only\ if,\ x \in A \\ 0\ if,\ and\ only\ if,\ x \notin A \\ 0 \leq \mu(x) \leq 1\ if\ x\ partial\ membership\ to\ A \end{cases} \tag{2}$$

The most evident characteristic of fuzzy logic is to consider that between two values (zero and one) there may be intermediate values, and these values are analyzed according to a degree of pertinence, which indicates the level that the information belongs to a specific set in a universe of discourse, according to [44].

Fuzzy set theory provides a method for manipulating sets whose boundaries are imprecise rather than restricted. The uncertainty of an element, that is, its fractional degree of pertinence, can be conceived as a measure of possibility, in other words, the possibility that an element is a member of the set [42].

2.2.1 Fuzzy inference systems

In many practical systems, relevant information comes from two sources: human experts, who describe their knowledge about the system in natural languages, and sensory measures and mathematical models proposed according to physical laws. An important task, therefore, is to combine these two types of information into systems designs [45].

The fuzzy inference system consists of a fuzzification interface, a rule base, a database, a decision-making unit or inference unit, and finally a defuzzification interface. The functional blocks are shown in **Figure 4**.

The function of each block is:

- A rule base containing a number of "if-then" fuzzy rules;
- A database that defines the functions of association of fuzzy sets used in fuzzy rules;
- A decision unit that performs rule inference operations;
- A fuzzification interface that transforms crisp inputs into degrees of correspondence with linguistic values;
- A defuzzification interface that transforms the fuzzy results of the inference into a crisp output.

Based on natural language, a fuzzy logic system is simple to understand and enables the representation and processing of human knowledge in a computer. The inputs, outputs, and fuzzy logic rules are easy to modify. These fuzzy logic features make it particularly well suited for use in a decision support system and is able to assist in the construction of vague rate-based irrigation control maps based on results of an imaging system in real time or by prescriptive maps based on the soil-plant-atmosphere transfer.

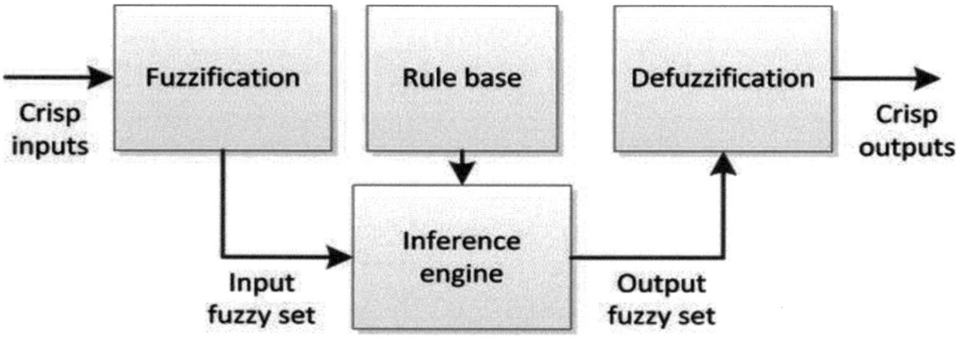

Figure 4.
Fuzzy inference system. Source: Adapted from Ross [46].

2.2.2 Mamdani inference method

Developed by Mamdani [47], the inference method is the most common in practice and literature. To begin the general view of this idea, it is considered a simple system of two rules, where each rule comprises two antecedents and one consequent. The graphic procedures herein illustrated can be easily extended and maintained for fuzzy rule bases or fuzzy systems with any number of antecedents and consequents. Two different cases of two-input Mamdani systems are considered, where the inputs to the system are scalar values and a max-min inference method is used. Thus, the Mamdani inference method for a set of conjunctive rules for r_{th} rules are given by Eq. (3):

$$if\ x_1\ is\ A_1^k\ and\ x_2\ is\ A_2^k\ then\ y^k\ is\ B^k\ for k = 1, 2, \ldots, r \quad (3)$$

This equation has a very simple graphical interpretation, exemplified in **Figure 5**, and illustrates graphical analysis of two rules, where symbols A11 and A12 refer to the first and second fuzzy antecedents of the first rule, respectively, and symbol B1 refers to the consequent fuzzy of the first rule. The symbols A21 and A22 refer to the first and second fuzzy antecedents, respectively, of the second rule, and the symbol B2 refers to the consequent fuzzy of the second rule.

2.2.3 Takagi-Sugeno-Kang inference method (TSK)

Although originally proposed by Takagi and Sugeno [48], this method is also known in the literature as Takagi-Sugeno-Kang (TSK) model. This is due to the subsequent works by Sugeno and Kang [49] related to methodologies developed to identify this type of model. The fuzzy TSK model consists of an inference system capable of describing, in an exact or approximate way, non-linear dynamic systems through a set of linear, locally valid dynamic systems, smoothly interpolated, non-linear and convex. A typical rule in a Sugeno model, which has two inputs, x and y, and one output z, is in the form of Eq. (4).

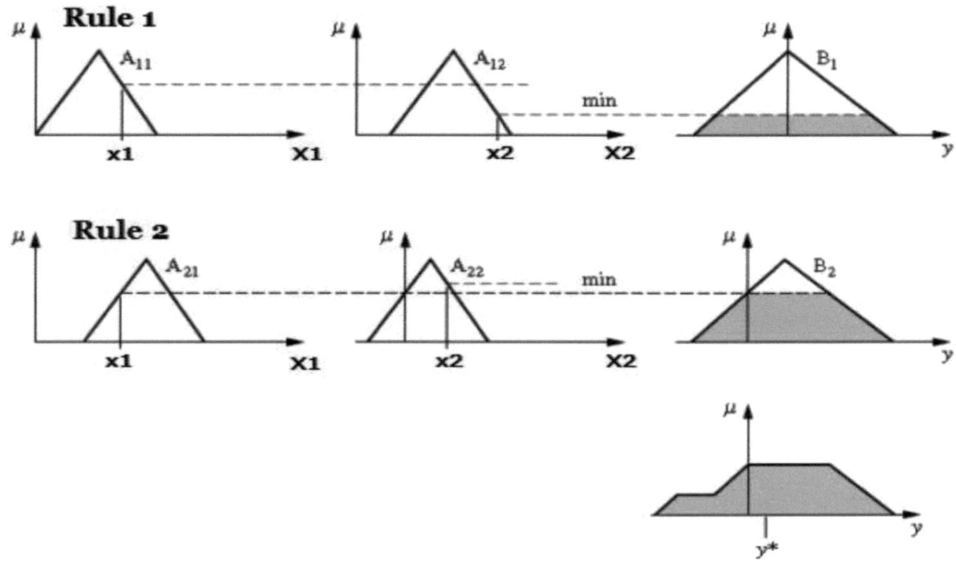

Figure 5.
Interpretation of the Mamdani method. Source: Adapted from Ross [46].

$$if\ x\ is\ A\ and\ y\ is\ B\ then\ z\ is\ z = f(x,y) \tag{4}$$

Usually, $f(x, y)$ is a polynomial function at the x and y inputs, but it can be any general function as long as it describes the output of the system within the fuzzy region specified in the antecedent of the rule to which it is applied. When $f(x,y)$ is a constant, the inference system is called a zero-order Sugeno model, which is a special case of the Mamdani system, in which the consequent of each rule is specified as a singleton fuzzy [50]. Each rule, in Sugeno model has an output given by a function. Due to this, the result is obtained through a weighted average, thus avoiding the time spent with the defuzzification process necessary in the Mamdani model. **Figure 6** illustrates the concept of the TSK model.

2.3 Remote sensing

Remote sensing technologies are being used more and more often in the precision agricultural applications. This is because that the variables (crop stress, soil type, disease,) to be measured and controlled are very disperse in remote areas with limited wireless communications or no power supply. Also, the measurements of each variable at spatial and temporal scale are expensive and time-consuming for installing and maintaining sensors over each field. Sensors can be multispectral cameras on Satellites or mounted on Unmanned Aerial Vehicle (UAV, or "drones"). In this chapter, we focalized on the using of satellites images for agricultural applications.

Remote sensing imagery can be used for mapping soil properties, classification of crop species (land use), detection of crop water stress, monitoring of irrigation, and predicting of crop yield. The use of remote sensing in precision agriculture depends principally on the spatial, temporal, radiometric and spectral resolution. Satellite remote sensing has shown a very strong potential for irrigation management at large scale through using a different data (optical, thermal and radar) acquired from different satellites.

Optical reflectances in red and near infrared (0.4–12.5 μm) have the potential to access the vegetation indices (VI) that are directly related the different crop parameters like crop coefficient (Kc) used in estimating the crop water requirements. Several studies (e.g., [51–59]) have been specifically dedicated for

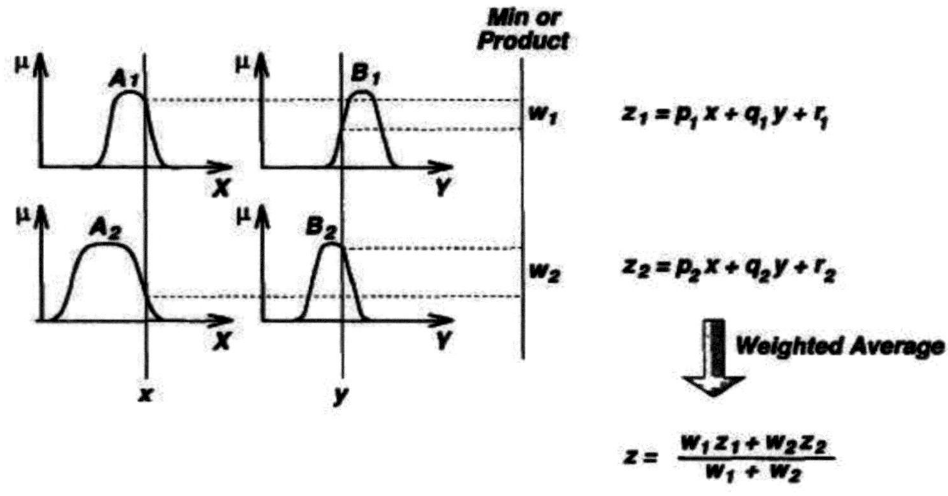

Figure 6.
Graphic interpretation of the Sugeno method. Source: Adapted from [50].

estimating Kc from Normalized Difference Vegetation Index NDVI [60] and Soil Adjusted Vegetation Index SAVI [61].

For thermal data, land surface temperature (LST) derived from thermal infrared remote sensing data have been used in a variety of applications such as, among others, climate studies [62, 63], the monitoring of crop water consumption and water stress detection [64–67], vegetation monitoring [68, 69], soil moisture estimation [70–72]. Canopy temperature has long been recognized as a good indicator for crop water status and as a potential tool for irrigation scheduling. Stomatal closure is one of the first responses to plant water stress that causes a decrease in plant transpiration and thus an increase in plant temperature. An increase in plant temperature is a sign that the vegetation is undergoing water stress. The crop water stress index (CWSI) is the most frequently used index to quantify the crop water stress based on canopy surface temperature [73].

Regarding the radar images, a significant effort has been recently dedicated to exploit these images to estimate soil moisture (SM) due to (i) the high-spatial resolution achievable by synthetic aperture radars (SAR) and (ii) the advent of SAR data available at high-temporal resolution. Especially, the Sentinel-1 (S1) constellation (composed of two satellites S1-A and S1-B) potentially provides SAR data at 20 m resolution every 3 days [74]. Thus, numerous studies have investigated and exploited the sensitivity of the radar signal to SM [70, 75–80].

3. Methods

The present stage shed light on the form that the construction of the proposed system is given, presenting as each fundamental characteristic of an irrigation project is appropriate to be added to the basic elements of fuzzy system.

The development of the intelligent irrigation system follows the structure shown in **Figure** 7. The structure of the proposed system enables the elaboration of a systematic, autonomous and automated management map to control an irrigation system. The output values of the intelligent system will be the inputs of the central pivot movement speed control and the sprinkle valve opening control.

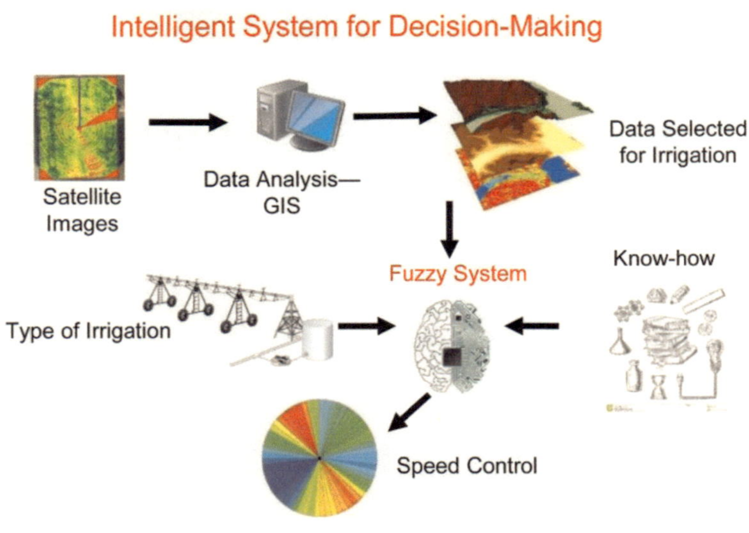

Figure 7.
Structure for intelligent irrigation system strategy.

However, it is important to emphasize that the commercial systems most used by farmers are not yet capable of elaborating this type of control map in the same way proposed by this work.

3.1 Geographic information system

Over the last decade, new information technologies, such as the geographic positioning system (GPS) and the geographic information system (GIS) have been introduced, which enabled to reduce the scale of management to the field level [81]. There are different software programs available in the market that can create maps from data point files, such as Surfer (GoldenSoftware, Inc.), ArcView (ESRI) and Global Mapper (Global Mapper).

The free QGIS[1] software will be used in this work for pre-processing and editing the file provided by the i-ekbase web-tool. QGIS is an open source geographic information system (GIS), licensed under GNU General Public License. It is an official open source geospatial foundation (OSGeo) project that runs on Linux, Unix, Mac OSX, Windows and Android, and supports several formats of vectors, rasters, databases and functionalities. QGIS has a plug-in infrastructure, and it is possible to add new features by writing plug-ins in C++ or Python.

3.1.1 Vegetation indices

As mentioned above, vegetation indices generated from remote sensing data are an important tool for the monitoring of natural or anthropogenic changes in land use and land cover. These rates have been used to estimate several vegetation parameters such as leaf area index (LAI) and amount of green biomass, as well as in the evaluation of land use and management and the recovery of degraded areas [82].

In this study, satellite image information was used, and in this case, the reading values of NDVI—Normalized Difference Vegetation Index—is defined as the difference between Near infra-red and red reflecatnces divided by its sum. It measures the vegetative cover and its color on the land surface over wide areas. Dense and green vegetation absorbs strongly the red wavelengths of sunlight and reflect in the near-infrared wavelengths resulting high values of NDVI, near to 1. For bare soil (no vegetation), NDVI values are between 0 and 0.14 depending on the moisture and roughness of soil. The practice of plant irrigation management has inherent complexity in visualizing the symptoms of water deficit, which are difficult to detect. On some occasions, they are discovered very late, that is, when observed, their effects have already compromised the production or quality of the product. Usually these symptoms are related to leaves coloring, leaf winding, leaf angle, etc.

However, it is possible to establish a correlation between the values of NDVI and the crop coefficient (Kc), [83, 84]. The estimated Kc values (Kc-NDVI) and the Kc values observed in Allen [85] for maize and soybean crops to guide the irrigation schedule during the season. Another way of relating the development of the plantation by means of remote sensing is the use of canopy temperature and infrared thermometry. A plant under water stress reduces transpiration and typically presents a higher temperature than the non-stressed crop [86], which can be a powerful tool for monitoring and quantifying water stress.

Canopy temperature increases when solar radiation is absorbed [87] but, is cooled when latent energy or sweating is used to evaporate water instead of heating plant surfaces. Algorithms based on canopy temperature are strongly correlated

---

[1] http://www.qgis.org/en/site/

with quantifiable crop yields [88], such as productivity, water use efficiency, seasonal evapotranspiration, leaf water potential at noon time, irrigation rates and damage caused by herbicides.

3.1.2 Satellite images

In order to study satellite images, data will be provided by a specialized company, by means of its intelligent environmental knowledgebase (i-ekbase), and made available via web tool, with limited and free use, for research related to the topic. The web tool will provide data from the area chosen initially for the study. The intelligent environmental knowledgebase (i-ekbase)[2] is an autonomous Big Data Analytics engine with a CLOUD system, and a fully automated geographic information system (GIS) [89]. **Figure 8** illustrates an example image provided by the i-ekbase tool, while **Table 1** shows the data generated by the web tool in the CSV (Comma Separated Values) format.

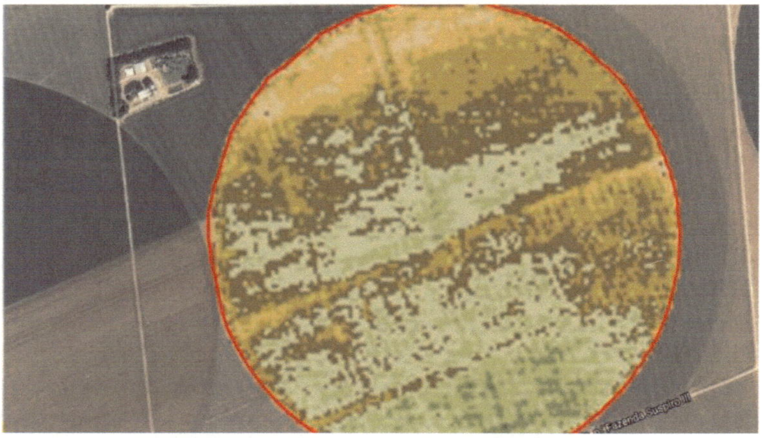

Figure 8.
Land surface temperature image by the i-ekbase web tool. Source: Adapted from the i-ekbase system.

Lat	Long	Canopy nitrogen (%)	Leaf area index (m^2/m^2	NDVI (%)	Bio-mass (tn/ha	Soil salinity (dS/m)	Soil moisture (%)	Canopy temp. (°C)
−15.2464	−54.0157	0.0	0.0	13.49	0.0	3.35	13.52	36.48
−15.2464	−54.0156	0.09	0.0	15.24	0.14	3.32	13.19	36.81
−15.2464	−54.0155	0.41	0.0	15.36	0.15	3.39	13.93	36.07
−15.2464	−54.0159	3.36	0.0	22.76	0.76	3.16	11.61	38.39
−15.2464	−54.0158	4.96	0.0	26.68	1.09	3.10	11.00	39.00
−15.2463	−54.0162	7.37	0.0	31.78	1.52	2.87	8.65	41.35
−15.2463	−54.0162	9.30	1.0	36.34	1.89	2.80	8.03	38.97
−15.2463	−54.0161	11.59	1.0	41.42	2.32	2.68	6.84	40.16

Table 1.
Data exported by the i-ekbase web tool. Source: Adapted from the i-ekbase system.

2 http://iekbase.com/

The i-ekbase system services provide larges area-wise resource management maps, with supporting remote digital scouting for decision support systems and rapid intervention of issues. For developing the experimental system were processed 12 months of Data, these remote sensing imageries were acquired by Landsat (with a spatial resolution of 30 m, but for this experiment the Data was upscale to 10 m) and Sentinel (with a spatial resolution of 10 m) satellites. Data that constitute this image have more than 14,000 georeferenced points, containing at each point or pixel the attributes of the agricultural analysis. Due to the extension of the data, only a few lines are shown in **Table 1**.

In order to apply this approach to the commercial field scale, the remote sensing data required to describe the soil-plant-atmosphere relationship can be acquired from satellite [90] and aircraft images [91, 92]. However, high costs, spatial resolution, data frequency and data availability [93, 94], in addition to cloudless satellite imagery, are a challenge for the correct execution of models based on remote sensing [95]. These issues can limit the efficiency of real-time variable rate irrigation management.

From the remote sensing data, those that best describe the soil-plant-atmosphere relationship for the intelligent irrigation system of the plantation site will be selected. In this phase, the correct selection of these data is fundamental to correctly calculate the results. A simple but promising approach uses crop coefficients derived from the normalized difference vegetation index (NDVI) along with local climate data to infer quantities of evapotranspiration (ETc) from variable crops almost in real time [57, 83, 96].

Based on the choice of planting site and type of crop to be irrigated, in relation to plant type data, the crop coefficient will be used along with information from the satellite images. In this case, the reading values of NDVI, near soil moisture and vegetative canopy temperature will be used. The latter is an important parameter for irrigation management and should be adjusted according to local growing conditions.

3.2 Location of the study

The study site is a farm located in the municipality of Primavera do Leste, MT, latitude 15° 14′24.73 "S and longitude 54 ° 0'53.29" W. This site has areas of cultivation irrigated by central pivot, and the crops planted are soybean, cotton and second-crop corn. The delimited area presents a total of 140 ha, in a radius of 667 m, see **Figure 9**. The area delimited by the red circle has central pivot irrigation, and the information used in the case study is from a 2015/2016 second-crop corn cycle. Irrigation in maize crop means to meet the minimum water requirements for the development of the crop.

Maize expresses high sensitivity to droughts. Therefore, the incidence of periods with reduction of the water supply to the plants at critical moments of the development of the crop, from flowering to physiological ripeness, can cause a direct reduction in the final harvest. In order to obtain maximum output, maize planting requires approximately 650 mm of water during its cycle [97], which can vary from 110 to 140 days in medium-cycle hybrids. For this preliminary analysis, data on daily average precipitation were used, provided by INMET (National Institute of Meteorology), from April to September 2016, to the city of Primavera do Leste, in the State of Mato Grosso, Brazil. **Figure 10** shows the data obtained.

These readings recorded during the development of the plantation under study corroborate the supposition of water stress due to lack of rainfall (from June to September), which would indicate the possibility of complementing water demand by irrigation.

Figure 9.
RGB images of the location of study area in in the municipality of primavera do Leste, state of Mato Grosso, Brazil.

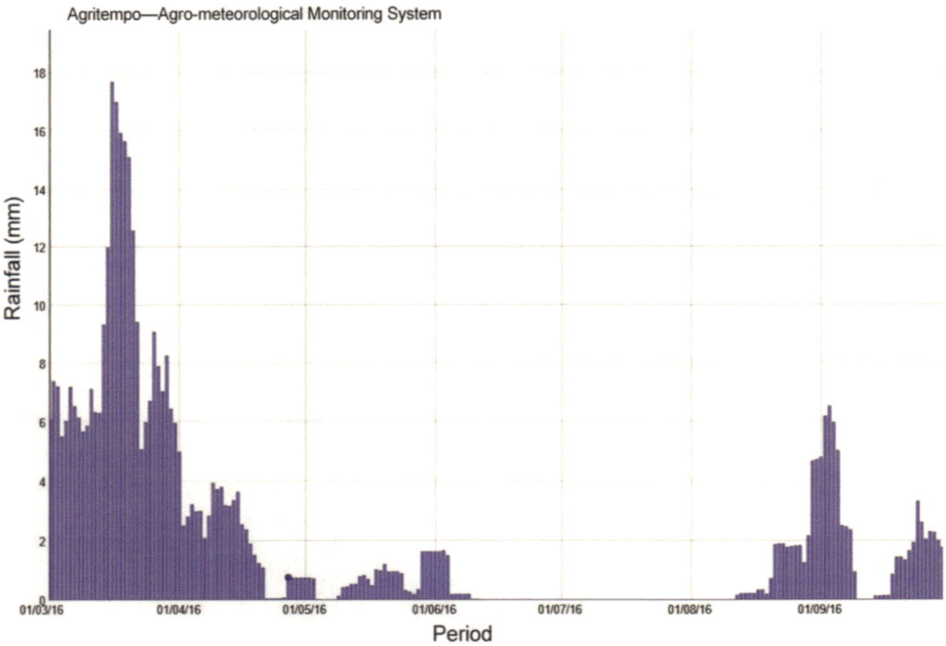

Figure 10.
Daily average precipitations obtained in the period of 2016. Fonte: INMET.

3.3 Fuzzy systems

In this step, a fuzzy system will be used, which in this case will be capable to infer the variations of linear speeds of the pivot according to the images provided by the satellite. For the creation of the control map, a system with artificial intelligence will be developed, capable of manipulating data and knowledge.

Three input variables (NDVI, near-soil moisture and canopy temperature) were used to infer the speed that the pivot should have to improve the level of irrigation within the management area, so that an adequate speed could be found for the

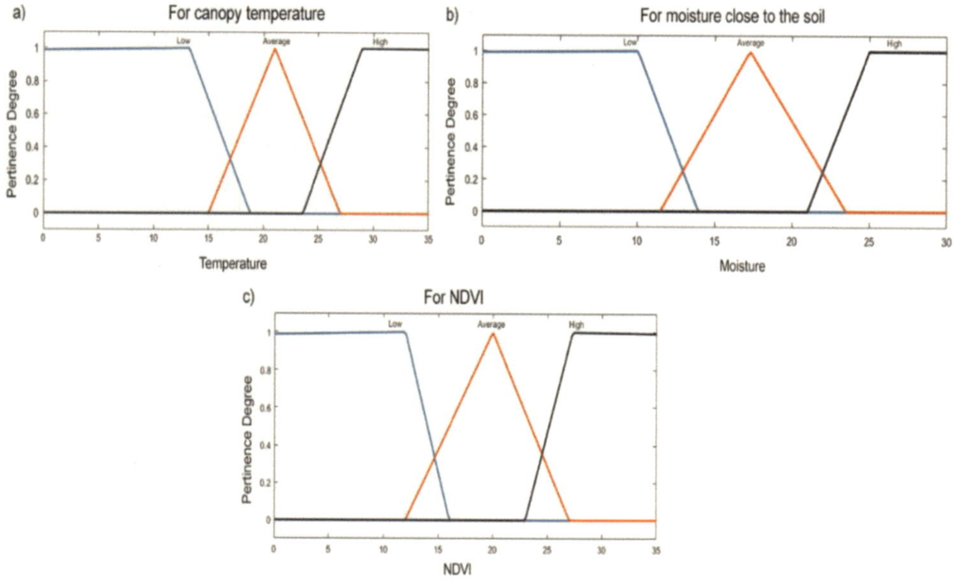

Figure 11.
Corresponding membership functions for each system entry, (a) canopy temperature, (b) upper layer soil moisture, (c) NDVI.

Input variables	Linguistic variables		
	Low	Average	High
Canopy temperature (°C)	<14	14 < ϕ < 27	>24
Upper layer soil moisture (%)	<14	12 < ϕ < 24	>21
NDVI (%)	<16	12 < ϕ < 27	>27

Table 2.
Fuzzy input set for the fuzzy inference system.

movement of the pivot in relation to the amount of water sprayed by the sprinklers. The decision unit or inference machine to perform rule-based inference operations will be implemented using the Mamdani method, with crisp inputs and crisp output value[3].

In this first stage of development, the water depth that the irrigation system provides will be considered constant, and the database, which defines the functions of association of the sets used in the fuzzy rules, will be implemented as shown in **Table 2** and **Figure 11**.

With the remote sensing data, it is possible to construct the universes of discourse of each input variable and thus transform the database into linguistic variables, such as those presented in the table above. Each of these inputs was previously limited in the universe of discourse in question and associated with a degree of pertinence in each fuzzy set by means of specialist knowledge. In this manner, in order to obtain the degree of pertinence of a given crisp input, it is necessary to search for this value in the knowledge base of the fuzzy system. The fuzzification of the decision-making system is shown in **Figure 11**, and it is possible

[3] https://www.agritempo.gov.br/agritempo/jsp/Grafico/graficoMicrorregiao.jsp?siglaUF=MT

to visualize the corresponding membership functions, considering these intervals as the universe of discourse of these variables.

Triangular membership functions were chosen because they simplify the calculation of the fuzzy inference mechanism. Well distributed triangular membership functions transform the input values into fuzzy values (low, medium and high), as shown in **Figure 5**, as well as the values of soil moisture and NDVI (**Figure 11b** and **c**, respectively). The fuzzy output set, which represents the rotational speed of the central pivot, was built on five linguistic variables: very low (VL), low (L), normal (N), high (H) and very high (VH). These sets were interpreted by means of their degrees of pertinence, illustrated in **Figure 12**.

If the center of gravity method is used for defuzzification, the fuzzy set produced after aggregation will be a numerical output composed of the union of all rule contributions. This calculation is made according to Eq. (5):

$$\mu^* = \frac{\sum_{i=1}^{n} \mu_i \bullet \mu_{out}(\mu_i)}{\sum_{i=1}^{n} \mu_{out}(\mu_i)} \tag{5}$$

The values $\mu_{out}(\mu_i)$ represent the area of a pertinence function modified by the result of fuzzy inference, and (μ_i) is the position of the centroid of the individual pertinence function.

Finally, the basis of fuzzy rules IF-THEN was elaborated and presented in **Table 3**, the fuzzy rule relating to rotation speed contains 27 rules, thus, the Mamdani inference method for a set of conjunctive rules is given by Eq. (3), for example: IF NDVI is Low AND Canopy temperature is Low AND Near-soil moisture is Low THEN Rotation Speed is Low.

This set of rules is based on the basic knowledge about irrigation, according to a methodology adopted by [37, 39].

The rules were constructed with the connective "AND", and are based on the supposition that where there is little leaf growth, there is soil water deficit. Together with the characteristic of the high canopy temperature, indicating a lower evapotranspiration, that is, water stress of the plants, the values of near-soil moisture provided by the web tool are readings of the locations where there are few leaves, and it is possible to estimate their value.

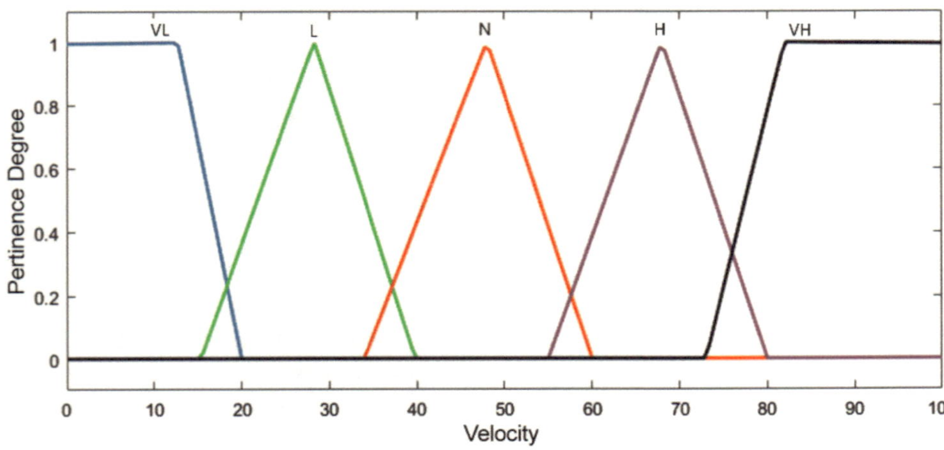

Figure 12.
Function of pertinence of the speed corresponding to the defuzzification of the system.

Inputs			**Output**
NDVI	**Temperature**	**Near-soil moisture**	**Rotation speed**
Low	Low	Low	Low
		Medium	Low
		High	Very low
	Medium	Low	Low
		Medium	Low
		High	Very low
	High	Low	Low
		Medium	Low
		High	Very low
Medium	Low	Low	Normal
		Medium	Normal
		High	High
	Medium	Low	Normal
		Medium	Normal
		High	High
	High	Low	Low
		Medium	Normal
		High	Normal
High	Low	Low	High
		Medium	Very high
		High	Very high
	Medium	Low	Very high
		Medium	Very high
		High	Very high
	High	Low	Very high
		Medium	Very high
		High	Very high

Table 3.
Fuzzy rules for central pivot speed control.

4. Control maps of central pivot motion speed

The development of the crop is evidenced captured by the images throughout the crop, and the information contained in this sensing is the NDVI values. By analyzing the information contained in **Figure 13**, it is possible to verify the similarity between the values attributed to Kc. It is noticed that as the crop evolves, the greater the exposure of the leaf area, and thus it is possible to establish a relation of NDVI. This process is described in [56], with ratios to calculate the base crop coefficient (Kcb) for cotton as a function of NDVI. When we look closely at each stage of the development of the plantation, two distinct areas are noticed: one with

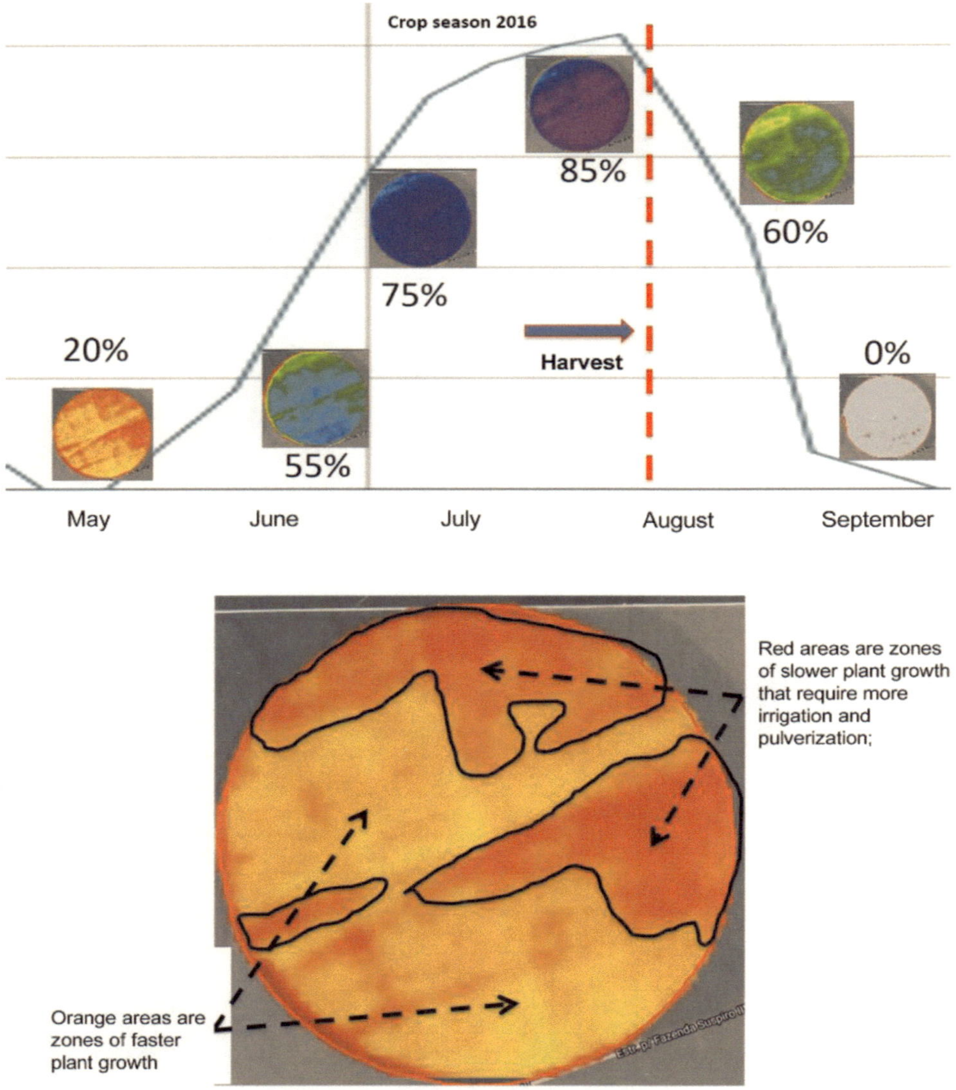

Figure 13.
Variation of NDVI in one crop cycle.

little growth and the other with normal growth. From this type of differentiation, it is possible to construct water demand maps, as well as speed control maps.

Data contained in this remote sensing are described in **Table 4**. In this configuration, the table presents the preprocessed data, near-soil temperature, soil moisture, and NDVI, besides latitude and longitude.

4.1 Case study on June 15th, 2016

Canopy temperature, near-soil moisture and NDVI data, analyzed and processed, will be the set of inputs for the intelligent fuzzy system. The following are illustrated in **Figure 14**: (a) NDVI images, (b) temperature images, and (c) soil moisture images.

The inputs, as shown in **Figure 14**, are arranged according to the linguistic variables of the fuzzy system and separated by tonalities for better visualization.

Lat	Long	NDVI (%)	Near-soil moisture (%)	Canopy temperature (°C)
−15.2463	−54.0157	5.96	25.75	30.75
−15.2463	−54.0156	6.49	25.68	30.24
−15.2463	−54.0154	6.67	25.68	30.03
−15.2463	−54.0153	6.85	25.68	30.19
−15.2463	−54.0152	6.66	25.8	30.63
−15.2463	−54.015	6.47	25.92	30.67
−15.2463	−54.0149	6.82	25.84	30.44
−15.2463	−54.0142	7.01	25.77	29.33
−15.2463	−54.0141	6.37	25.88	29.58

Table 4.
Pre-processed data.

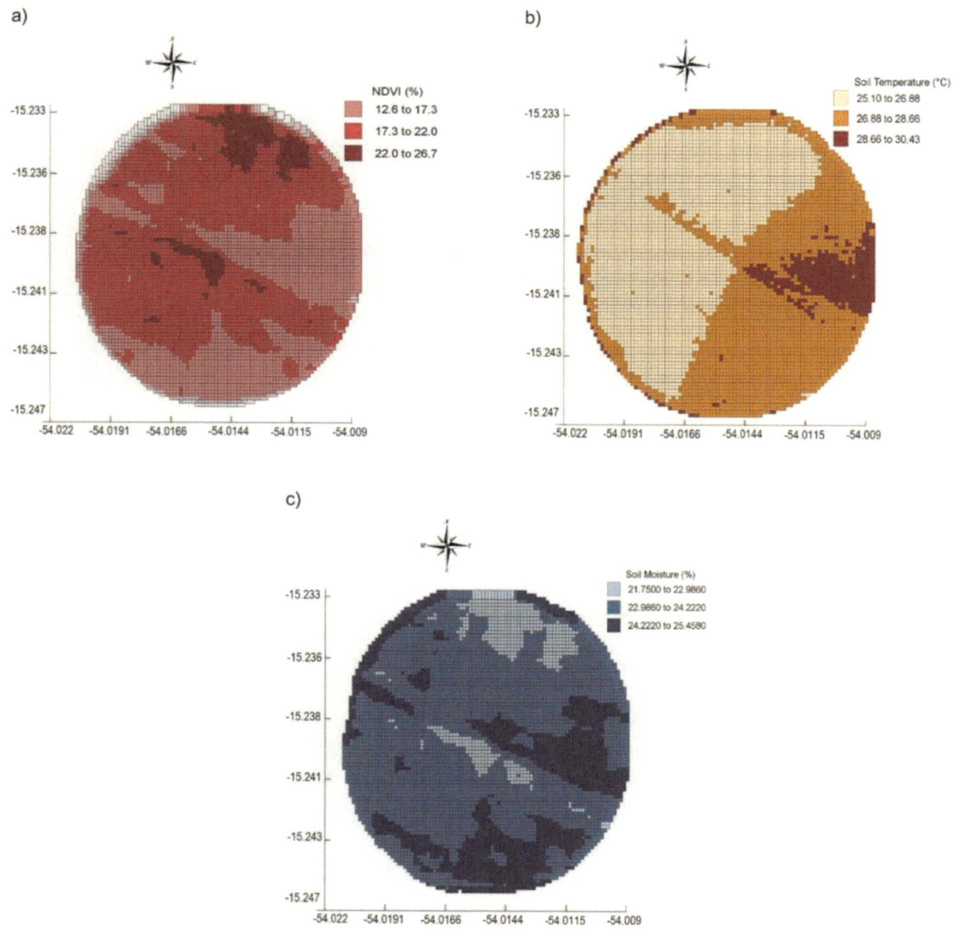

Figure 14.
Input data from the fuzzy inference system, (a) NDVI data, (b) canopy temperature data, (c) near-soil moisture data.

The intelligent system gave the result shown in **Figure 15**, where it is possible to verify different regions within the area, with different values for the pivot rotation speed. The indirect relationship between the pivot rotation speed and the level of the applied water depth implies a smaller applied water depth in a higher speed, and

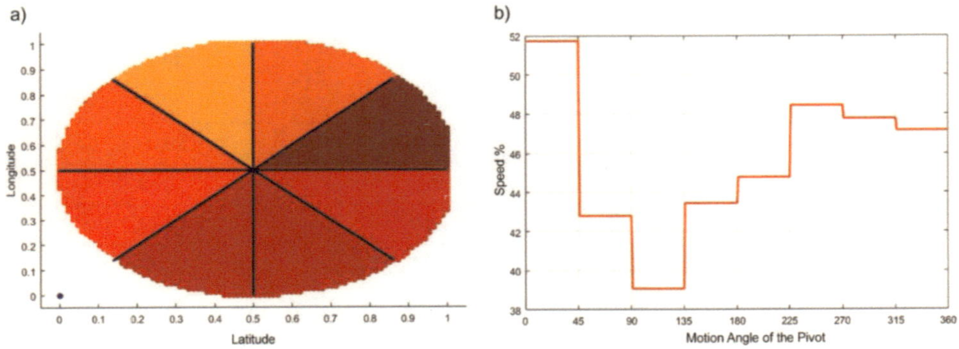

Figure 15.
Control map of pivot rotating speed, (a) speed control map, (b) pivot turning speed setpoints.

a higher water application in the soil in a lower rotation speed [98]. When analyzing the input data, it is possible to identify two large areas with a lower leaf development, which may indicate a lack of water for development. After processing this input data, the intelligent irrigation system indicates that these areas with lower leaf development, in a redder color, indicate that the pivot should reduce its speed and thus increase the water depth in that area.

The expected result is the creation of control maps, and in this case, it was possible to determine the speed reference values for the eight zones initially programmed. The areas that presented different coloration in **Figure 14** are in the control map result. It is possible to verify well divided zones, and in each one there is a determined value for the speed that the pivot must develop to decrease or increase the water depth in the cropped area. The result shown in **Figure 15b** corresponds to the reference values that should be sent to the pivot controller, since the control systems of these devices work with percentage of rotation speed.

4.2 Case study on June 28th, 2016

The data analyzed and processed by the GIS were used as inputs to the intelligent fuzzy system. They are illustrated in **Figure 16**: (a) NDVI images, (b) temperature images and (c) soil moisture images.

Similar to the previous case study, the study of June 28 presents the values of the input variables of the fuzzy system with the linguistic definitions necessary for interpretation. The results of the intelligent irrigation system are shown in **Figure 17**, where is also possible to observe different regions within the crop area, with different values for the pivot rotation speed. A higher speed of rotation implies a smaller applied water depth, and with a lower speed of rotation, there is a greater application of water to the soil, if the application flow is kept constant by the sprinklers.

When comparing satellite images once again, it is seen that NDVI and canopy temperature are essential for the decision-making of the intelligent irrigation system. It is possible to see that there are large areas with a lower leaf development, which may indicate a lack of water for development. In the case of intelligent irrigation system output, areas in a redder color indicate that the pivot should slow down.

The expected result is the creation of the control maps, and for this study it was possible to find the reference values of the central pivot rotation speed for the eight irrigation zones initially programmed, shown in **Figure 17**. In this result, it is also possible to identify the areas that presented different colors in **Figure 16**.

DOI: http://dx.doi.org/10.5772/intechopen.87023

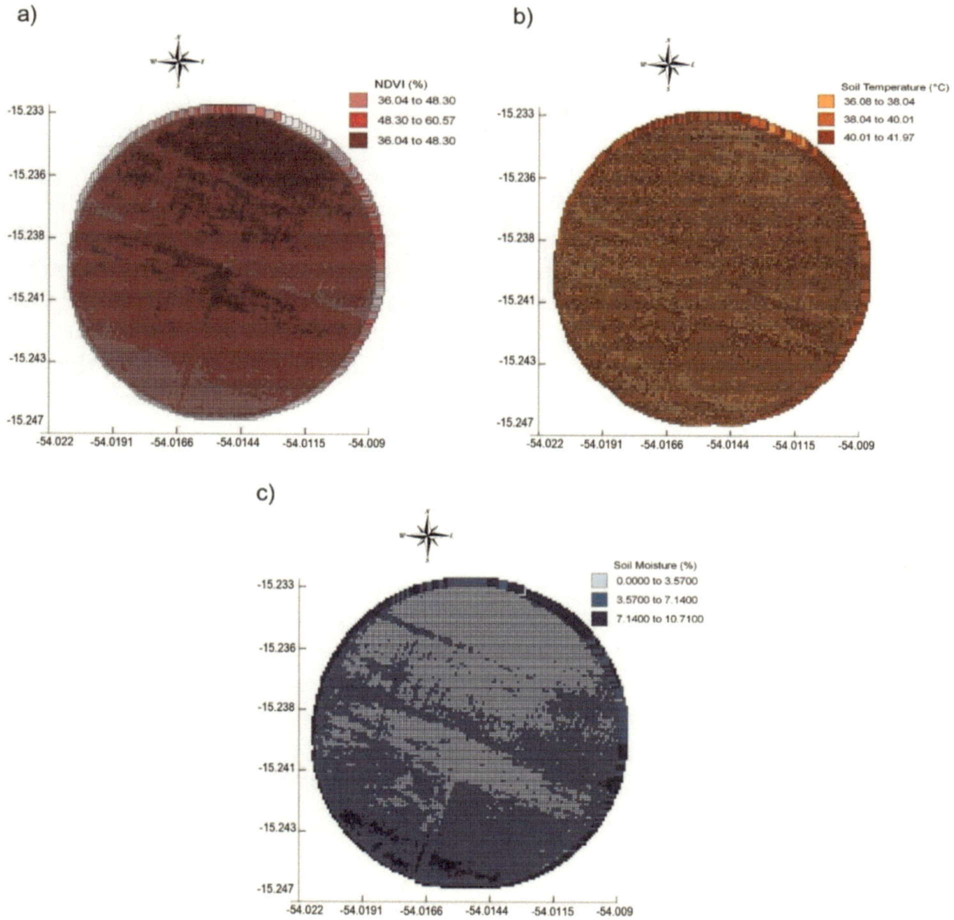

Figure 16.
Input data from the fuzzy inference system, (a) NDVI data, (b) canopy temperature data, c) near-soil moisture.

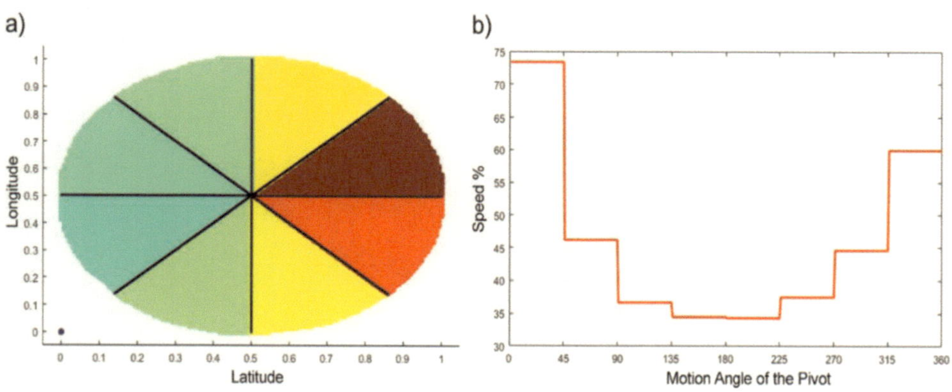

Figure 17.
Pivot rotating speed control map, (a) speed control map, (b) pivot rotation speed setpoints.

The irrigation management zones are fairly divided, and in each one a value is determined for the pivot rotation speed, decreasing or increasing the water depth applied to the crop area. The result in **Figure 17b** corresponds to the reference values to be sent to the pivot controller.

5. Conclusions

The fuzzy control system for irrigation developed is original and ground breaking, and there is no literature about a rotation speed control map for center pivot in the same approach presented in this work. In addition, there is no available information that commercial systems can build this type of map autonomously.

The experiments point to the efficiency result for pivot operation, since it is possible to note differences between the speeds per management zone that could be employed in the pivot. The system follows the definition of variable rate irrigation, since when changing the speed, the amount of water applied also changes.

In this context, it was observed that the fuzzy logic can be widely used in farming, and it is feasible to aggregate precision irrigation knowledge with the formulation of a decision support system. The implementation was successful for the application of variable rate water to central pivots. However, a broader commercial application depends on the integration of data collection systems, management strategies, and hardware control.

These studies were motivated by a broader research effort on the applications of fuzzy systems in agriculture. In addition, fuzzy control applied in variable rate irrigation (VRI) was explored in this domain in order to provide a better understanding of the relation between agricultural factors involving complexity and uncertainty and solutions with A.I. technologies. Future development and application of these methodologies in agricultural engineering are required especially in the context of decision support in precision irrigation. The results are favorable to the continuity of the studies on precision irrigation and application of the fuzzy logic for the development of control maps for central pivots irrigation systems.

Acknowledgements

The authors thank the Federal Institute of Education, Science and Technology of Mato Grosso—IFMT—for the financial support, and to the Post-Graduation Program in Electrical and Computer Engineering of the Federal University of Rio Grande do Norte—PPGEEC/UFRN—for the technical and administrative support. Finally, we thank the Coordination of Improvement of Higher Education Personnel —Ministry of Education (Capes—MEC) for granting the PRODOUTORAL scholarship to the corresponding author.

Conflict of interest

The authors declare no conflict of interest.

Author details

Willians Ribeiro Mendes[1]*, Fábio Meneghetti U. Araújo[2] and Salah Er-Raki[3,4]

1 Control Engineering and Automation Department, Federal Institute of Education, Science and Technology of Mato Grosso—IFMT, Mato Grosso, Brazil

2 Computer Engineering and Automation Department, Federal University of Rio Grande do Norte, University Campus Lagoa Nova, Natal, Rio Grande do Norte, Brazil

3 LP2M2E, Département de Physique Appliquée, Faculté des Sciences et Senvolvimen Techniques, Université Cadi Ayyad, Marrakech, Morocco

4 Center for Remote Sensing Application (CRSA), University Mohammed VI Polytechnic (UM6P), Benguerir, Morocco

*Address all correspondence to: willians.mendes@pdl.ifmt.edu.br

References

[1] ANA. Agência Nacional de Águas (Brasil). Levantamento da Agricultura Irrigada Por Pivôs Centrais no Brasil-2014: Relatório síntese/Agência Nacional de Águas. Vol. il. Brasília: ANA; 2016. p. 33. ISBN: 978-85-8210-034-9

[2] FAO. The State of Food and Agriculture. Climate Change, Agriculture and Food Securty. Rome: Food and Agriculture Organization of the United Nations; 2016

[3] Christofidis D. Água na produção de Alimentos: O Papel da irrigação no Alcance Do Desenvolvimento sustentável. Brasília: Universidade de Brasília; 2005. p. 29

[4] Balastreire LA. A Estado-da-Arte da Agricultura de Precisão no Brasil. Ed. Vol. 1. Piracicaba; 2000

[5] Brase T. Precision Agriculture. Boston, United States: Cengage Delmar Learning Publishers Inc; 2005:288. ISBN-13: 978-1-40188-105-4

[6] Srinivasan A. Handbook of Precision Agriculture: Principles and Applications. Binghamton, NY: CRC Press Taylor & Francis Group; 2006. ISBN-13: 978-1-56022-954-4

[7] Bernardi AC. Agricultura de precisão: Resultados de Um Novo Olhar. Brasília-DF: Embrapa; 2014. 596 p. Available from: http://ainfo.cnptia.embrapa.br/digital/bitstream/item/113993/1/Agricultura-de-precisao-2014.pdf. ISBN 978-85-7035-352-8

[8] Gilley JR, Mielke LN, Wilhelm NN. An experimental center-pivot irrigation system for reduced energy crop production studies. Transactions of ASAE. 1983;**26**(5):1375-1379

[9] Bright JC. Optimal control of irrigation system: An analysis of water Allocation Rules [thesis]. University of Canterbury: Department of Agricultural Engineering; 1986

[10] Fraisse CW, Heermann DF, Duke HR. Modified linear move system for experimental water application. In: Advances in Planning, Design, and Management of Irrigation Systems as Related to Sustainable Land Use. Vol. 1. Lueven, Belgium: Center for Irrigation Engineering, K. U. Leuven; 1993. pp. 367-376

[11] Sadler EJ, Camp CR, Evans DE, Usrey LJ. A site-specific center pivot irrigation system for highly-variable Coastal Plain soils. In: Robert PC, Rust RH, Larson WE, editors. Proceedings of the Third International Conference on Precision Agriculture. Madison, Wis: Am. Soc. of Agron; 1996. pp. 827-834

[12] Evans RG. If you Build it, Will it Work? Testing Precision Center-Pivot Irrigation. GPS World Magazine, Eugene; 1997. pp. 10-17

[13] Evans RG, King BA. Site-specific sprinkler irrigation in a water-limited future. Transactions of the ASABE–Advances in Irrigation. 2012;**55**:493-504

[14] McCarthy AC, Hancock NH, Raine SR. Advanced process control of irrigation: The current state and an analysis to aid future development. In: Irrigation Science. Vol. 31. Springer-Verlag; 2013. pp. 183-192. DOI: 10.1007/s00271-011-0313-1

[15] Er-Raki S, Chehbouni A, Duchemin B. Combining satellite remote sensing data with the FAO-56 dual approach for water use mapping in irrigated wheat fields of a semi-arid region. Remote Sensing. 2010;**2**(1):375-387

[16] Amazirh A, Er-Raki S, Chehbouni A, Rivalland V, Diarra A, Khabba S, et al. Modified penman–Monteith

equation for monitoring evapotranspiration of wheat crop: Relationship between the surface resistance and remotely sensed stress index. Biosystems Engineering. 2017; **164**:68-84. DOI: 10.1016/j.biosystemseng.2017.09.015

[17] Deguise JC, Mcnairn H. Hyperspectral remote sensing for precision agriculture. In: Proceedings of Fifth International Conference on Precision Agriculture (CD); 16–19 July; Bloomington: MN, USA; 2000

[18] McCarthy AC, Hancock NH, Raine SR. Real-time data requirements for model-based adaptive control of irrigation scheduling in cotton. Australian Journal of Multi-Disciplinary Engineering. 2011;**8**(2):189-206. ISSN 1448-8388

[19] Boutraa T, Akhkha A, Alshuaibi A, Atta R. Evaluation of the effectiveness of an automated irrigation system using wheat crops. Agriculture and Biology Journal of North America. 2011;**2**(1): 80-88

[20] Caceres R, Casadesus J, Marfa O. Adaptation of an automatic irrigation control tray system for outdoor nurseries. Biosystems Engineering. 2007;**96**:419-425

[21] Miranda FR, Yoder RE, Wilkerson JB, Odhiambo LO. An autonomous controller for site-specific management of fixed irrigation systems. Computers and Electronics in Agriculture. 2005;**48**: 183-197

[22] Luthra SK, Kaledhonkar MJ, Singh OP, Tyagi NK. Design and development of an auto irrigation system. Agricultural Water Management. 1997; **33**:169-181

[23] Colaizzi PD, Evett SR, Agam N, Schwartz RC, Kustas WP, Cosh MH, et al. Soil heat flux calculation for sunlit and shaded surfaces under row crops: 2. Model test. Agricultural and Forest Meteorology. 2016:129-140. DOI: 10.1016/j.agrformet.2015.10.009

[24] Anom SW, Shibusawa S, Sao A, Sakai K, Sato H, Hirako S, et al. Soil parameter maps using the real-time spectrophotometer. In: Proceedings of Fifth International Conference on Precision Agriculture (CD); 16–19 July; Bloomington: MN, USA. 2000

[25] Dubois D, Prade H. Fuzzy Sets and Systems: Theory and Applications. New York: Academic Press; 1980

[26] Rommelfanger HJ. Fuzzy logic-based processing of expert rules used for checking the credit solvency of small business firms or for supporting analytic procedures of auditors. In: Ribeiro RR, editor. Soft Computing in Financial Engineering. Vol. 37. Heidelberg: Physica-Verlag; 1999. pp. 1-244

[27] Dubey S, Pandey RK, Gautam SS. Literature review on fuzzy expert system in agriculture. International Journal of Soft Computing and Engineering (IJSCE). 2013;**2**(6)

[28] Zhang HY, Fei L, Wei Z, Congcong M, Yuewei C. The fuzzy decision-making method of irrigation amount based on ET and soil water potentioal. In: International Conference on Electronics, Communications and Control (ICECC); 2011. 2927-2931

[29] Werner A, Jarfe A. Preagro: Precision Agriculture, Herausforderung an Integrative Forschung, Entwicklung Und Anwendung in der Praxis. Proceedings 2002. Darmstadt, Germany: Kuratorium für Technik und Bauwesen in der Landwirtschaft e.V. (KTBL); ISBN: 3-9808279-0-9

[30] Domsch H, Giebel A. Electrical conductivity of soils typical for the state of Brandenburg in Germany. In: 3rd European Conf. on Precision

Agriculture; 18–20 June; Montpelier: France. 2001. pp. 373-378

[31] Sparovek G, Schnug E. Soil tillage and precision agriculture: A theoretical case study for erosion control in Brazilian sugarcane production. Soil and Tillage Research. 2001;**61**(1–2):47-54

[32] Heermann DF, Wallender WW, Bos MG. Irrigation efficiency and uniformity. In: Hoffman GJ, Howell TA, Solomon KH, editors. Management of Farm Irrigation Systems. 2nd print. Michigan: St. Joseph, ASAE; 1992. pp. 125-149

[33] Mulla DJ, Schepers JS. Key processes and properties for site-specific soil and crop management. In: Pierce FJ, Sadler EJ, editors. The State of Site-Specific Management for Agriculture. Madison, WI: ASA, CSSA, SSSA; 1997. pp. 1-18

[34] Lowenberg-Deboer J, Swinton SM. Economics of site-specific management in agronomic crops. In: Pierce FJ, Sadler EJ, editors. The State of Site-Specific Management for Agriculture. Madison WI: ASACSSA-SSSA; 1997. pp. 369-396

[35] Torre-Neto A, Schueller JK, Haman DZ. Networked sensing and valve actuation for spatially variable microsprinkler irrigation. In: Presented at the July 2000 ASAE International Meeting; Milwaukee, WI: Paper No. 001158. ASAE; 2950 Niles Rd. St. Joseph, MI; 2000

[36] Grisso R, Alley M, McClellan P, Brann D, Donohue S. Precision Farming: A Comprehensive Approach. Virginia Cooperative Extension; 2009. Available from: http://pubs.ext.vt.edu/442/442–500/442–500.html

[37] Bernardo S, Soares AA, Mantovani EC. Manual de Irrigação. 8a edição ed. Viçosa: Imprensa Universitária; 2006. p. 625

[38] da Silva JGF, Coelho EF. Irrigação do mamoeiro. In: Martins D, Dos S, Costa A, de F, da S, editors. A Cultura Do Mamoeiro: Tecnologia de produção. Vitória, ES: INCAPER; 2003. pp. 161-197

[39] da Silva EM, de Azevedo JA. Dimensionamento da Lateral de irrigação Do pivô Central. Planaltina: EMBRAPA-CPAC; 1998. p. 54

[40] Reichardt K. A Água Em Sistemas Agrícolas. Piracicaba: Editora Manole; 1990. p. 188

[41] Zadeh LA. Fuzzy sets. Information and Control. 1965;**8**:338-353

[42] Simões MG, Shaw IS. Controle e Modelagem Fuzzy. 2a ed. São Paulo: Editora Edgard Blucher Ltda; 2007

[43] Dubois D, Ostasiewicz W, Prade H. Fuzzy Sets: History and Basic Notions. In: Dubois D, Prade H, editors. Fundamentals of Fuzzy Sets. The Handbooks of Fuzzy Sets Series. Boston, MA: Springer; 2000;7:21-124. https://doi.org/10.1007/978-1-4615-4429-6_2

[44] Barros LC, Bassanezi RC. Tópicos de Lógica Fuzzy Com Aplicações Em Biomatemática. Campinas, SP: UNICAMP/IMECC; 2006

[45] Henderson H. Encyclopedia of computer science and technology. Facts on File. Rev. ed., New York; 2009. pp. 203-204. ISBN-13: 978-0-8160-6382-6

[46] Ross TJ. Fuzzy Logic with Engineering Applications. Handbook. 2nd.Ed ed. USA: John Wiley & Sons, Ltd. University of New Mexico; 2004. p. 628

[47] Mamdani EH, Assilian S. An experiment in linguistic synthesis with a fuzzy logic controller. International Journal of Man-Machine Studies. 1975;7: 1-13

[48] TAKAGI T, SUGENO M. Fuzzy identification of systems and its applications to modeling and control.

IEEE Transactions on Systems, Man, and Cybernetics. 1985;**SMC-15**(1): 116-132

[49] Sugeno ME, Kang GT. Structure identification of fuzzy model. Fuzzy Sets and Systems. 1986;**28**:329-346

[50] Jang J-SR, Sun C-T, Mizutani E. Neuro-Fuzzy and Soft Computing: A Computational Approach to Learning and Machine Intelligence. Prentice Hall; 1997

[51] Jackson RD, Idso SB, Reginato RJ, Pinter PJ Jr. Remotely sensed crop temperatures and reflectances as inputs to irrigation scheduling. In: Irrigation and Drainage Special Conference Proceedings; 23–25 July. Boise: Idaho, ASCE New York; 1980. pp. 390-397

[52] Bausch WC, Neale CMU. Crop coefficients derived from reflected canopy radiation: A concept. Transactions of ASAE. 1987;**30**(3): 703-709

[53] Neale CMU, Bausch WC, Heerman DF. Development of reflectance-based crop coefficients for corn. Transactions of ASAE. 1989;**32**(6):1891-1899

[54] Choudhury BJ, Ahmed NU, Idso SB, Reginato RJ, Daughtry CST. Relations between evaporation coefficients and vegetation indices studies by model simulations. Remote Sensing of Environment. 1994;**50**:1-17

[55] Hunsaker DJ, Pinter PJ Jr, Barnes EM, Kimball BA. Estimating cotton evapotranspiration crop coefficients with a multispectral vegetation index. Irrigation Science. 2003;**22**:95-104

[56] Hunsaker DJ, Pinter PJ Jr, Kimball BA. Wheat basal crop coefficients determined by normalized difference vegetation index. Irrigation Science. 2005;**24**:1-14

[57] Er-Raki S, Chehbouni A, Guemouria N, Duchemin B, Ezzahar J, Hadria R. Combining FAO-56 model and ground-based remote sensing to estimate water consumptions of wheat crops in a semi-arid region. Agricultural Water Management. 2007;**87**:41-54

[58] Er-raki S, Rodriguez JC, Garatuza-Payan J, Watts CJ, Chehbouni A. Determination of crop evapotranspiration of table grapes in a semi-arid region of Northwest Mexico using multi-spectral vegetation index. Agricultural Water Management. 2013; **122**:12-19. DOI: 10.1016/j.agwat.2013.02.007

[59] Diarra A, Jarlan L, Er-Raki S, Le Page M, Aouade G, Tavernier A, et al. Performance of the two-source energy budget (TSEB) model for the monitoring of evapotranspiration over irrigated annual crops in North Africa. Agricultural Water Management. 2017; **193**:71-88. DOI: 10.1016/j.agwat.2017.08.007

[60] Rouse JW, Haas RH, Schell JA, Deering DW, Harlan JC. Monitoring the Vernal Advancement and Retrogradation of Natural Vegetation. Greenbelt MD: NASA/GSFC, Type III, Final report; 1974. pp. 1-371

[61] Huete AR. A soil-adjusted vegetation index (SAVI). Remote Sensing of Environment. 1988;**25**: 295-309

[62] Hansen J, Ruedy R, Sato M, Lo K. Global surface temperature change. Reviews of Geophysics. 2010; **48**(4):1-29. DOI: 10.1029/2010RG000345

[63] Kustas W, Anderson M. Advances in thermal infrared remote sensing for land surface modeling. Agricultural and Forest Meteorology. 2009;**149**: 2071-2081. DOI: 10.1016/j.agrformet.2009.05.016

[64] Bastiaanssen WGM, Menenti M, Feddes RA, Holtslag AAM. A remote sensing surface energy balance

algorithm for land (SEBAL). Journal of Hydrology. 1998;**212-213**:198-212

[65] Boulet G, Chehbouni A, Gentine P, Duchemin B, Ezzahar J, Hadria R. Monitoring water stress using time series of observed to unstressed surface temperature difference. Agricultural and Forest Meteorology. 2007;**146**: 159-172. DOI: 10.1016/j.agrformet. 2007.05.012

[66] Er-Raki S, Amazirh A, Ayyoub A, Khabba S, Merlin O, Ezzahar J, et al. Integrating thermal surface temperature into penman-Monteith model for estimating evapotranspiration and crop water stress of orange orchard in semi-arid region. Acta Horticulturae. 2018; **208**:120-131. DOI: 10.17660/ActaHortic. 2018.1197.12

[67] Olivera-Guerra L, Merlin O, Er-Raki S, Khabba S, Escorihuela MJ. Estimating the water budget components of irrigated crops: Combining the FAO-56 dual crop coefficient with surface temperature and vegetation index data. Agricultural Water Management. 2018; **208**:120-131. DOI: 10.1016/j. agwat.2018.06.014

[68] Kogan FN. Operational space technology for global vegetation assessment. Bulletin of the American Meteorological Society. 2001;**82**: 1949-1964. DOI: 10.1175/15200477 (2001)082<1949:OSTFGV>2.3.CO;2

[69] Williamson HD. Evaluation of middle and thermal infrared radiance in indices used to estimate GLAI. International Journal of Remote Sensing. 1988;**9**:275-283. DOI: 10.1080/ 01431168808954851

[70] Amazirh A, Merlin O, Er-raki S, Gao Q, Rivalland V, Malbeteau Y, et al. Retrieving surface soil moisture at high spatio-temporal resolution from a synergy between Sentinel-1 radar and Landsat thermal data: A study case over bare soil. Remote Sensing of Environment. 2018;**211**:321-337. DOI: 10.1016/j.rse.2018.04.013

[71] Merlin O, Duchemin B, Hagolle O, Jacob F, Coudert B, Chehbouni G, et al. Disaggregation of MODIS surface temperature over an agricultural area using a time series of Formosat-2 images. Remote Sensing of Environment. 2010;**114**:2500-2512. DOI: 10.1016/j.rse.2010.05.025

[72] Sandholt I, Rasmussen K, Andersen J. A simple interpretation of the surface temperature/vegetation index space for assessment of surface moisture status. Remote Sensing of Environment. 2002; **79**:213-224. DOI: 10.1016/S0034-4257 (01)00274-7

[73] Moran MS, Clarke TR, Inoue Y, Vidal A. Estimating crop water deficit using the relation between surface-air temperature and spectral vegetation index. Remote Sensing of Environment. 1994;**49**:246-263. DOI: 10.1016/0034-4257(94)90020-5

[74] Torres R, Snoeij P, Geudtner D, Bibby D, Davidson M, Attema E, et al. GMES Sentinel-1 mission. Remote Sensing of Environment. 2012;**120**:9-24. DOI: 10.1016/j.rse.2011.05.028

[75] Baghdadi N, Gaultier S, King C. Retrieving surface roughness and soil moisture from synthetic aperture radar (SAR) data using neural networks. Canadian Journal of Remote Sensing. 2002;**28**(5):701-711. DOI: 10.5589/ m02-066

[76] Baghdadi N, Holah N, Zribi M. Soil moisture estimation using multi-incidence and multi- polarization ASAR data. International Journal of Remote Sensing. 2006;**27**:1907-1920. DOI: 10.1080/01431160500239032

[77] Holah N, Baghdadi N, Zribi M, Bruand A, King C. Potential of ASAR/ ENVISAT for the characterization of soil surface parameters over bare

agricultural fields. Remote Sensing of Environment. 2005;**96**(1):78-86. DOI: 10.1016/j.rse.2005.01.008

[78] Srivastava HS, Patel P, Manchanda ML, Adiga S. Use of multiincidence angle RADARSAT-1 SAR data to incorporate the effect of surface roughness in soil moisture estimation. IEEE Transactions on Geoscience and Remote Sensing. 2003. DOI: 10.1109/TGRS.2003.813356

[79] Zribi M, Baghdadi N, Holah N, Fafin O. New methodology for soil surface moisture estimation and its application to ENVISAT-ASAR multi-incidence data inversion. Remote Sensing of Environment. 2005;**96**(3–4):485-496. DOI: 10.1016/j.rse.2005.04.005

[80] Vereecken H, Huisman JA, Bogena H, Vanderborght J, Vrugt JA, Hopmans JW. On the value of soil moisture measurements in vadose zone hydrology: A review. Water Resources Research. 2008;**44**:1-21. DOI: 10.1029/2008WR006829

[81] Blackmore BS, Griepentrog HW. A future view of precision farming. In: Proc. PreAgro Precision Agriculture Conf. Muncheberg, Gemrnay: Centre for Agricultural Landscape and Land Use Research (ZALF); 2002. pp. 131-145

[82] Okin GS. Relative spectral mixture analysis: A multitemporal index of total vegetation cover. Remote Sensing of Environment. 2007;**106**:467-479. DOI: 10.1016/j.rse.2006.09.018. Toulouse

[83] González-Dugo MP, Mateos L. Spectral vegetation indices for benchmarking water productivity of irrigated cotton and sugarbeet crops. Agricultural Water Management. 2008; **95**:48-58

[84] Kamble B, Kilic A, Hubbard K. Estimating crop coefficients using remote sensing-based vegetation index. Remote Sensing. 2013;**5**:1588-1602. New York

[85] Allen RG, Pereira LS, Raes D, Smith M. Crop Evapotranspiration-Guidelines for Computing Crop Water Requirements, Irrigation and Drain, Paper No. 56. Rome, Italy: FAO; 1998. p. 300

[86] Bellvert J, Zarco-Tejada PJ, Girona J, Fereres E. Mapping crop water stress index in a "pinot-noir" vineyard: Comparing ground measurements with thermal remote sensing imagery from an unmanned aerial vehicle. Precision Agriculture. 2014;**15**:361-376

[87] Idso SB, Baker DG. Relative importance of reradiation, convection and transpiration in heat transfer from plants. Plant Physiology. 1967;**42**: 631-640

[88] Colaizzi PD, O'shaughnessy SA, Evett SR, Howell TA. Using plant canopy temperature to improve irrigated crop management. In: 24th Annual Central Plains Irrigation Conference. Colby KS: CPIA; 2012

[89] Dutta R, Morshed A, Aryal J, D'este C, Das A. Development of an intelligent environmental knowledge system for sustainable agricultural decision support. Environmental Modelling & Software. 2014;**52**:264-272

[90] Moran MS, Inoue Y, Barnes EM. Opportunities and limitations for image-based remote sensing in precision crop management. Remote Sensing of Environment. 1997;**61**:319-346

[91] Wood GA, Taylor JC, Godwin RJ. Calibration methodology for mapping within-field crop variability using remote sensing. Biosystems Engineering. 2003;**84**:409-423

[92] Fitzgerald GJ, Lesch SM, Barnes EM, Luckett WE. Directed sampling using remote sensing with a response

surface sampling design for site-specific agriculture. Computers and Electronics in Agriculture. 2006;**53**:98-112

[93] Pinter PJ Jr, Hatfield JL, Schepers JS, Barnes EM, Moran MS, Daughtry CS, et al. Remote sensing for crop management. Photogrammetric Engineering & Remote Sensing. 2003;**69**:647-664

[94] Trout TJ, Johnson LF, Gartung J. Remote sensing of canopy cover in horticultural crops. HortScience. 2008; **43**:333-337

[95] Barker JB, Heeren DM, Neale CMU, Rudnick DR. Evaluation of variable rate irrigation using a remote-sensing-based model. Agricultural Water Management. 2018;**203**:63-74

[96] Hunsaker DJ, Barnes EM, Clarke TR, Fitzgerald GJ, Pinter PJ Jr. Cotton irrigation scheduling using remotely sensed and FAO-56 basal crop coefficients. Transactions of ASAE. 2005;**48**:1395-1407

[97] Bergamaschi H, Dalmago GA, Bergonci JI, Bianchi CAM, Muller AG, Comiran F, et al. Distribuição hídrica no período crítico do milho e produção de grãos. Pesquisa Agropecuária Brasileira, Brasília. 2004;**39**:831-839

[98] Valín MI, Cameira MR, Teodoro PR, Pereira LS. DEPIVOT: A model for center-pivot design and evaluation. Computers and Electronics in Agriculture. 2012;**87**:159-170

Chapter 4

Performance of Water Desalination and Modern Irrigation Systems for Improving Water Productivity

Hani Abdelghani Mansour, Ren Hongjouan, Hu Jiandong, Bao Hong Feng and Liang Changmei

Abstract

Desalination is the process that is performed to remove excess salts from water to become potable or agriculture. This applied science is now concerned by many countries suffering from water shortage. Over the next ten years, this science is expected to grow significantly due to the expected water crises in many countries. The consumption of energy in the desalination process is one of the important problems and difficult obstacles that need to be overcome. The Egyptian water strategy should include increasing amount of desalinated water to more than 50%, especially since Egypt is in a very rich location in saltwater sources and they can be utilized to the maximum extent possible. The researchers have attempted to develop varieties of some traditional crops such as wheat, saline resistant to salinity using local selective ecotourism techniques and using genetic engineering through which saline-tolerant genes are added, but it can be said that so far these efforts have not resulted in the production of candidate seawater breeds The maximum salinity of irrigation water in the long term, even for the most salt-tolerant crops such as date palm, is still less than 5 mmol.

Keywords: saline water, irrigation system, water management, desalination, magnetic water

1. Introduction

1.1 Salt farming toward a greener future

Saline farming is based on the cultivation of crops and plant varieties that can tolerate high levels of salinity and temperature. Its idea originated primarily from nature itself and from the growth of naturally salty plants on sandy beaches, tidal areas, saline lands, and other saline-flooded areas [1–4]. The most prominent example of this type of plant is the mangrove plant, also known as the Crimea or Shura, which is heavily distributed on the shores of the Red Sea and the shores of the Arabian Gulf. The Crimea—and this type of plant in general—adapts to the saline environment in more than one way and mechanism. Salts, which are collected and then disposed or adapted to the saline environment in the place they are found, contain a high degree of salinity [5, 6]. Some researchers have tried to mimic these natural conditions by using saline water to irrigate and grow some plant and crop species. One may succeed in adapting to high salinity and continuing to grow and produce [7]. The first attempt began in 1949, but the expansion of research and scientific experimentation on the cultivation of saline plants began only in the late seventies of the last century. Since then, many scientific institutions, such as Arab research centers, have been actively trying to develop new techniques for salt farming and to develop new varieties of salt-tolerant plants and crops, whether from major food crops such as wheat and barley or from other plants that can be exploited as natural pastures or fodder for livestock [8–10]. However, the concept of saline agriculture is not only to improve the ability of some plants to grow and mature in a harsh saline environment and to irrigate and plant certain plant and crop species with high salinity water but also to use brine or improve some of their properties or specifications, such as sugar concentration in fruit [11–14].

In general, it is possible to confirm that the process of developing saline farming techniques and the production of saline-resistant plant species took two different scientific approaches: the first was the application of genetic engineering technique to genetically and geologically transform traditional plants and transform them from being saline to tolerant. While the second approach is based on the cultivation of saline plants that can tolerate salinity and try to expand their cultivation in the wild and agricultural fields for use as food crops or animal feed or for the production of oilseeds.

The expansion of saline agriculture can be of greater benefit, such as the exploitation of arid and semi-arid lands, which sometimes have abundant amounts of brackish water unsuitable for conventional agriculture. Cultivating some suitable crop types and increasing their production will help to achieve food security [15–17]. Similarly, marginal beaches exposed to tidal movement can also be exploited for feed farming or other plant species that can be used to produce energy (biofuels) or to extract pharmaceuticals or oils [18–20].

Salt farming can also contribute to increasing the efficiency of the use of water resources by conserving potable water or traditional agriculture, which in turn helps to achieve water security and reduce migrations and conflicts resulting from the lack of water, land degradation, and increased drought. In addition, salt farming techniques can also contribute to improving the productive efficiency of some crops or improving the quality and characteristics of certain crops, thereby achieving high economic returns [21–24]. Saline farming can also contribute to mitigating the effects of global warming leading to climate change. This is due to its role in increasing agricultural land and green plant areas that absorb carbon dioxide from the atmosphere [25–29].

As such, salt farming can support traditional agriculture and raise pressure in it, since the requirement of traditional agriculture is to find less water-consuming ways and increase agricultural production to meet the growing demand for food and staple crops. Traditional agricultural activities are the main consumers of drinking water by 70%, followed by industrial activities by 20%, and other activities.

Pibars and Mansour [30] found that severe soil water deficit (SWD) decreased grain yield of winter wheat, while slight SWD throughout the growing season did not reduce grain yield or water productivity. This result indicates that water supply can be reduced somewhat without significant decrease in grain yield. Moreover, investigations conducted by [31, 32] show that deficit irrigation can increase the net farm income. Barley, considered as a tolerant plant [33], occupies large cultivated areas in arid parts of Tunisia. Many experiments have been conducted on barley cultivated in small private farms in southern Tunisia [34] and the results demonstrate the potential of irrigation management practices in reducing the effects of salinity on both yield and soil salinity. In addition, [35] showed that yield reduction under deficit irrigation during the whole growing season was about 5% and 20% of the total irrigation water was saved.

1.2 The need to use seawater in irrigation

There are about 295 million hectares of coastal desert land in the world, where about 17% (about 50 million hectares) is level 0% slope lands, it's suitable for irrigated agriculture (in terms of soil type, the land slope, and there is no competition for the other uses), and it's expected that the land will increase the irrigated areas in the desert regions by about 80%. The common regions for such use are the different deltas of some rivers where coastal sediments constitute sedimentary desert lands such as the Nile River (Egypt), Euphrates River (Iraq), and Colorado River (US), where the sedimentary coastal delta is often suffering from secondary salinity problems. Many places suffer from desertification [36, 37].

Such spaces can be introduced into economically valuable agricultural production by cultivating halophytes using seawater. The sandy coastal desert along the coast of the Red Sea and the Arabian Gulf, as well as the Indian Ocean and the Gulf of California, are suitable for this type of use, adding additional areas suitable for irrigation by seawater. Many coastal plains of the sea, as well as in the northern coastal plains of Australia and some areas close to population centers or some large cities such as Cairo, Baghdad, Bombay, and Karachi, are also common for such use. This is a great opportunity to invest in the production of animal feed from halophytes, which reduces the pressure on the use of fresh water and available agricultural land and reduces overgrazing on relatively few grasslands [38, 39].

It was projected that the desalination technology would provide nuclear power, where it could provide a cheap source of energy for seawater desalination that could be used for agricultural and land reclamation. But to some extent, these techniques are still expensive and require high investment. Therefore, the direct use of seawater in agriculture is an optimistic and great hope for agricultural development along the coastal deserts by growing salinity-rich and economically profitable crops [11–13].

Some researchers have tried to cultivate traditional crops on seawater such as barley, which can at least complete their life cycle on seawater in temperate climates [40–42]. Furthermore, it has been assumed that breeding programs for such crops can be developed to improve their salinity tolerance and to create mutations of resistance or salinity. But so far, no conventional crops have been found that can produce an acceptable economic yield under irrigation conditions in the sea in the coastal desert climate [43]. Recently, a new approach has been proposed: to attempt to settle or rehabilitate and cultivate saline-loving plants that grow naturally in such conditions for agricultural production and thus can be used as a natural resistance to salinity. Some countries, such as North Africa, have begun using this technique to produce seeds and halophytes using seawater. About 15 years ago, the University of Arizona began field experiments to cultivate halophytes in many parts of the desert world [44].

With increased experience and information, experimental plots were increased from 0.5–1 ha to 20–40 ha experimental farms, and different irrigation methods ranging from conventional surface irrigation to pivotal irrigation, which is about 250 ha, were used. [45].

Halophytes were selected to produce seeds from which oil is extracted, known as *Salicornia bigelovii torr*. Other varieties, such as Atriplex, longitudinal shrubs, and saline and other types of succulent plants, were tested for salinity. It was enough for the growth of most halophytes, which were salinity tolerant; nevertheless, the growth rate of these plants decreased by 50% due to irrigation by sea water.

However, in field trials, the results showed that the annual yield and seed yield may be equal to or greater than the yield of any traditional crops irrigated with fresh water. The biomass crop can be produced in the range of 17–34 ton/ha, which contains 11–23 ton/ha organic matter from halophytes and using seawater in field experiments for 6 years. An annual oil crop of 2 ton/ha of oilseed crop (equivalent to soybeans or other oil crops) was obtained [46–49].

The reason for the highest ability of halophytes to produce yields despite the deleterious effect of salinity of seawater is due to having many compensatory factors in such coastal spots where they do not compete with grasses or other pests and that halophytes have biological talent and ability that are high on photosynthesis and growth [22, 50–52]. In this chapter, we provide a revision about the possibilities of this new technology and how to assess the economic feasibility of different ways of using seawater in the production of animal feed. The final aim of the chapter is to find the suitable methods of water desalination and to use modern irrigation systems for the improvement of strategic crops in Egypt.

2. Materials and methods

2.1 Salinity of irrigation water and how to solve it

Drilling wells is the main problem when it comes to saline water and then it does not take long to find that it is salted, and hence the farms have a great role in the search with you for realistic solutions to address the salinity of irrigation water [53]. The different methods of desalination are as follows: (1) distillation by mixing fresh water with saline water, (2) method of magnetic water, and (3) using seawater in agriculture with high-tolerant plants.

2.1.1 Water salinity and agriculture

Water salinity is defined as the ratio of total soluble salts (sodium chloride, calcium sulfate, magnesium sulfate, and various bicarbonate salts) in 1 l of water [54].

2.1.2 Salinity of groundwater used in irrigation systems

Increasing the irrigation of saline groundwater reduces the strong attraction between the soil granules, thus reducing the interstitial spaces, causing difficulty in spreading the roots. This reduces the rate of root absorption of the water by the osmotic properties. The roots are burned, and the leaves of the plants and some branches are burned. Advanced sedimentation occurs when reverse osmosis occurs, causing the plant to die completely. In general, it is not correct to say that water salinity is treated, but it can be said that there is control over the effect of irrigation with saltwater [55]. Refs. [56–65] found the most important methods to control the salinity of irrigation with saltwater and to mitigate its harmful effects on plants (salinity treatment of water), which are mentioned in **Table 1**:

I	**Methods applied to the crop**
1	Cultivation of plants suitable to the degree of salinity of water, and this is the origin in the process of avoiding damage to the salinity of irrigation water.
2	When growing vegetable crops, the number of seedlings in acres or the quantity of seeds, to compensate for the loss of death or absence of germination or delay, must be increased.
3	It is recommended to spray micro-elements in paper in the form of Edita, and attention must be given to marine algae and amino acids to further encourage the plant to withstand the conditions that cause stress.
4	Potassium fertilization, especially in the form of potassium silicate, has an important role in resistance to salt stress and reduction of transpiration of the plant.
II	**Methods applied to the irrigation management and water sources**
1	Irrigation in the early morning and after sunset is preferred, because the sun rays lead to increase the proportion of water vapor from the surface of the soil and thus increase the accumulation of salts, and night irrigation causes the washing of salts and its interaction with the soil components of sand and clay and therefore less emphasis on the soil surface. It is then easy to wash with acid and saline treatment compounds.
2	Agriculture can be done in the upper third of the bottom lines to evaporate the floating of salts in the upper part of the line.
	The irrigation system should not be used permanently if the salinity ratio exceeds 1300 ppm. This is due to the rapid evaporation of the water when spraying with water, leaving the salts around the plant, causing severe damage, even death.
3	The drainage must be improved by continuous tillage to facilitate the washing of salts during heavy irrigation, especially the soil salts do not accumulate with water salts and cause serious damage.
4	The interval between the irrigation must be reduced to maintain high content of soil moisture.
5	Some farmers have recently moved to the work of a fish tank and duck before sacrificing water in irrigation pipes.
III	**Methods applied to the process of cultivation**
1	Adding the organic material before planting at a rate of 10 tons of compost per feddan works to hold water and transmit without evaporation and increase the permeability of the soil.
2	Turning the soil with hard plows must be avoided, especially in the desert lands, to avoid moving the salts from the depth of the soil to the top again.
3	Increasing the rates of fertilization with humic and fulvic ability to increase the capacity of cationic exchange of soil, and generally works on: PH regulation of the soil where hydroxic acid combines with calcium carbonate and produces free calcium and releases CO_2, which dissolves in water, forming a carbonic acid that makes the soil relatively acidic and increases its ability to absorb nutrients. Humic acid combines with sodium chloride, a component of sodium hydroxide, an organic complex that releases chlorine, a gas that inhibits pathogens in the soil.
4	Cultivation of windbreaks early because of their important role in reducing the speed of wind and reduce the temperature of soil, and with it increase the rate of water lag after irrigation and reduce the impact of harmful salts.

Table 1.
The most important methods to control the salinity of irrigation of saltwater and to mitigate its harmful effects on plants.

2.2 Solving the problem of soil salinity: how to reduce the salinity of underground water?

With the expansion of the reclamation of saline land, the increase in water scarcity, and the dependence of countries on well water, however, the problems of salinity have increased and have become even more threatening to agricultural production [66]. Investors began to look for ways to solve the problem of soil salinity, and researchers were interested in developing ways to reduce the proportion of salts in water [66–69]. In this report, we offer you several golden tips, which are the summary of experience and science in dealing with the problems of salinity in your farm or field: It is not wise for farmers in saline lands to wait to overcome the full salt problem before starting the commercial production cycle. Rather, it is wise to coexist with the problem and gradually overcome it by preparing the root region to coexist with the permitted limits of growth.

Soil permeability relationship with irrigation capacity of saline water: Land with good permeability is tolerated by irrigation with saline water up to 3000 ppm without the accumulation of salts or causing a problem [70]. Poor soil permeability is precipitated by water salts even if salinity is 200 ppm due to the accumulation of salts over time [71]. Calcium nitrate and urea were used as an alternative to nitrate as a source of nitrogen and calcium. Fertilizers containing sulfate, such as potassium sulfate, magnesium, manganese, and zinc, were used. Sulfuric acid is usually used through the irrigation operation by the (fertigation) techniques, whereas nitric and phosphoric acids are used with the organic matter during preparation for each new crop.

To reduce the harmful effect of salts in the well water, we must modify the ionic structure by adding some chemicals [72–74], which helps to precipitate the harmful constituents of carbonates and bicarbonates. The following relationships should also be known: sodium absorption ratio (SAR)/electrical conductivity (EC)/leaching ratio (LR), especially when using saline water, so that the soil does not deteriorate and decrease production. How to reduce the effect of water salinity on crops? To overcome the problem of salinity and achieve the highest productivity in the presence of salts, irrigation periods with the installation of long washing irrigation to remove the salts from the root area must be rounded. If the proportion of sodium in the irrigation water is to be considered, the approximate percentage of calcium or magnesium must be adjusted by adding calcium throughout the year.

The best source of calcium here is the agricultural gypsum because of its multiple benefits, which we will talk about in a separate report. Reducing salinity of wells by choosing irrigation method.

When sprinkling and high-water salinity, water droplets should be observed to be large and not misty. Irrigation is done at night so that the evaporation process decreases, and the salts are deposited on the leaves of the plant. Frequent drip irrigation without changing the lines or completely immersing the entire land once or twice a year will exacerbate the problem of salinity in the soil. In addition, irrigation of agricultural banks that have drainage water for farms and adjacent fields is the most dangerous to the future of the soil and will not be able to reduce the deterioration and desertification in the future. The large number of composts with fertilization of ammonium nitrate causes the salinity of the soil to be increased.

2.3 Salinity ratio suitable for agriculture

The problems of soil and water salinity are endless, and we have asked hundreds of questions about the best methods to be relied upon in agriculture. In the next report, we will review together important information on this subject, with an indication of the salinity ratio suitable for agriculture [75, 76]. Possible agricultural methods to avoid and reduce salinity damage based on nonreclaimed saline lands can be utilized as follows: Agriculture is on high lines with agriculture in the lower half of the miles because the salts bloom at its peak. The same method can be followed when farming on the terraces with the work of a small pyramid rise in the center of the terrace in order to bloom salts. Winter crops are preferred where the salt damage is less than that in the summer crops, and planting is preferred by seedling [57]. Drip irrigation helps to collect salts away from plants, so that the soil is washed from the accumulated salts before planting the next crop. Methods of solving the problem of salinity of the soil to obtain the salinity ratio suitable for agriculture: Salinity treatment is not easy, so it is necessary to control and coexist with salts in soil. So as not to exceed the limits allowed by the integration of agricultural operations of plowing, fertilization, irrigation, drainage and treatment of salinity by following this process (**Figure 1**).

Where soil samples are not analyzed, the following system shall be followed: The plowing of the earth is two orthogonal slits. To wash the soil, wash by irrigation by immersing or spraying with sprayers at a rate of 100 m^3 per feddan once a week. Soil salinity testing is recommended after each washing routine to determine the effect of washing on salinity and to know whether the washing process is continuing. Add 20 m^3 of my squash per feddan + 200 kg superphosphate—this is for growing vegetables. [45] In case of trees, 5 m^3 + 50 kg superphosphate is added to the line of agriculture only, and the mixture is divided into 60 cm depth and 80 cm width. It is preferable to sterilize organic compost by sun or use organic manure. The organic fertilizer is composted in one area and is well soaked to saturation and covered with plastic for 2–3 months. Fertilizer rates are added with the addition of appropriate washing requirements with a good drainage system.

2.4 How are the washing needs of the plants that are added to the irrigation water determined to help achieve the appropriate salinity ratio for the plant?

This is done by the following equation:

Laundry needs = salinity of irrigation water (mm) × (100/plant tolerance rate of salinity).

Example: The wet needs of potato plants irrigated with water with an electrical conductivity of 1 mm/cm at 25°C are required [42].

The ECe, where there is no crop shortage, is 1.7 millimhos.

Solution: washing needs of potato = 1 × (100/1.7) = 59%.

In order to avoid any shortage in the potato crop, which is infused with salted water 1 ml/cm, it is necessary to increase the amount of water needed for each irrigation [48]. This is 59% as washing needs to wash the accumulated salts in the area of root spread and drain away from the root zone.

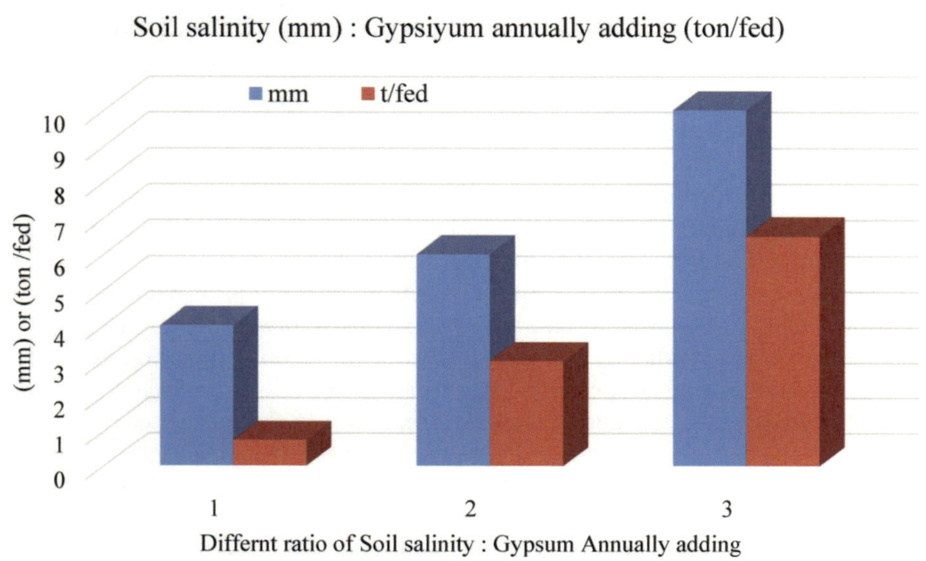

Figure 1.
The ratio between mean of soil salinity and amount of gypsum annually adding. (1) 4 mm: 0.75 ton/fed; (2) 6 mm: 3 ton/fed; (3) 10 mm: 6.5 ton/fed.

It is noted that these washing needs calculated in the example are very high and may not be followed, especially because of the lack of irrigation water or the absence of good drains [40].

Therefore, the tolerance of the plant for salinity is calculated on the basis of the ECe score, where a 25% reduction in yield occurs in case of potatoes at ECe = 3.8.

Thus, the washing requirements of potato = 1 × (100/3.8) = 26% more than the amount of water assessed for each irrigation as washing needs to wash the accumulated salts in the area of spreading the roots and away from them. Commercial sulfuric acid is injected with irrigation water at a rate of 2 l per feddan per week for a month. This removes the salts from the roots and removes them on the surface of the soil, thus improving the growth of the plants. Some natural compounds and raw materials are used for the treatment of salinity.

2.5 Magnetic water to overcome saltwater and make it suitable for agriculture

Many countries in the world, including the Arab countries, suffer from the loss of water of the river. Therefore, desalinated saltwater is available from seawater and groundwater. In fact, there are several techniques to desalinate salty water, including the following:

1. Chemical deposition using CaO lime

2. Distillation

3. Electrical switchboard

4. Ion exchangers

5. Membranes or so-called reverse osmosis

6. Evaporation and condensation by the rays of the sun

There is an excellent way to decompose using magnetic separation. This efficiently removes up to 99% and more. This is because of the salinity of the water (the presence of positive ions and other negative ions), but these ions do not attract toward the magnet, so scientists thought of the physical center attracted by the negative ions and positive ions and attracted them to the center of magnets and this is called Feret Fe_3O_4, which is thrown in the saline water to be reanalyzed and attracted to the negative and positive ions by the magnetic field, which penetrates water for purifying it. Electrical dialysis has been commercially known since the 1960s, 10 years before reverse osmosis. The cost is effective for desalination of saline water wells and makes the decision for attention in this regard. The electrolysis technique is based on the following general principles. The electrolysis unit consists of several hundred pairs of cells connected to each other by electrodes called a compound of membranes. Feeding water flows simultaneously through passages through the cells to provide the flow of desalinated water as the concentrated water passes through the compound.

Based on the design of the system, it is possible to add chemicals in the compound to reduce the voltage and prevent the formation of crusts. Feed water must be treated from the outset to prevent substances that sweat membranes or block narrow channels in cells from entering the membrane compound. Feed water is rotated through the compound with a low-pressure pump to overcome water resistance as it passes through narrow passages. A rectifier is often installed to convert the oscillating current into a direct current supplied to the electrodes from outside the membrane complexes. Final treatment includes water stabilization and processing for distribution, which may include the removal of gases such as hydrogen sulfide or alkaline modification.

2.6 Selection of soil preparation site

Land processing and settlement are important factors in irrigated agriculture, especially when using seawater in irrigation. As is known in irrigated agriculture, salts tend to accumulate and redistribute in the soil sector, where salinity occurs in the field. For example, high areas are increasingly accumulating salt. Therefore, the land should be divided into pieces that may be different in size, but attention should be paid to settling the soil surface in one piece.

Many soil species have been successfully used in clay land to sand dunes and it is important that the soil is good for natural drainage. It is therefore necessary to plow the soil to a depth of 1 m to improve drainage especially in heavy land. When sandy soil is compressed under the surface layer, it must be prepared in terms of deep plowing, surface tillage, settlement, agriculture, and irrigation.

Although many halophytes bear high ground water, interest in drainage is an important factor in resisting salinization. In the case of heavy land, shallow banks must be made in the form of a letter V, at a depth of half a meter, and in dimensions of 10–20 m, to be drained into deep drains and can be drained through water pumps back into the sea.

The surface of the desert sea has a shallow reservoir of groundwater saline that extends to several kilometers of sea level. However, irrigation by seawater will cause damage to any groundwater reservoir. Therefore, hydrological studies should be carried out for the aquifers of the area to be used for irrigation by seawater (depth—quality—quantity). The hydrophysical characteristics of the soil must be studied. If the site is located next to a mountain range parallel to the coast as in the case of the Red Sea, the fresh water that collects under the mountain valley or on the sandy shoreline must be maintained.

One of the most important restrictions on the use of seawater to produce halophytes is how to manage water. It is necessary to prevent the accumulation of salts in the rhizosphere. This is a condition other than freshwater irrigation where irrigation is based on the level of soil moisture. In traditional irrigation conditions, irrigation occurs when the soil moisture is reduced to 50%. However, in the case of seawater irrigation and the lack of ground moisture to 50%, the salinity level in the root zone is twice the salinity level of seawater, which has a severe effect on the plant. For most halophytes, it was found that the moisture deficiency should not exceed 25% to reduce the chance of increasing salt concentration between the irrigation in the soil sector. It is also necessary to add washing needs about 25% or more in each rye to wash the salts and expel them below the root area.

Short irrigation and high-salt washing are therefore key to achieving success and achieving high yields of halophytes using seawater. For example, in the case of sand dunes or sandy beaches, irrigation should be carried out regularly on a daily basis during the summer season, while in sandy soil, which can't retain enough water, irrigation can be done every day and every 10 days in the winter season.

2.7 Planning the establishment of a seawater farm

The most important factor required for the establishment of a seawater farm is that there is a source available, close to seawater and at a low cost. The cost of seawater supply is the largest investment in this type of project, and it exceeds the other factors such as irrigation method, quantity of water required, and agricultural practices required.

Usually, in the case of direct supply of seawater, it is a sea pier extending into the sea where pipes are drawn to fetch water by pumps. Irrigation channels must be constructed in the fields of the project, all of which affect the coast in terms of appearance and other uses of the coast. The movement of water and various marine organisms and the properties and effect of seawater on the rust of metals used in these marine environmental installations and the movement of waves, winds, and hurricanes are many difficult problems that must be considered and taken into account when designing. Solutions are usually expensive.

The alternative approach is indirect supply through wells to collect seawater, thus avoiding many of the above problems. Therefore, in the case of a groundwater reservoir, seawater wells are the best solution, but the limited capacity of the well may be a problem (many of these wells have been discovered despite their presence on the seashore in many studies). After a source of seawater is found, the next task is to connect the water to the root area, which can take many forms according to the irrigation method used. In the case of small spaces that can be used, simple irrigation method of watering where a system can be characterized by the rapid flow of water in open channels or light PVC pipes or plastic tubes can be folded. In the case of larger areas, the sprinklers can be used with either the axial or the lateral spray where the water is distributed homogenously on the ground even if the ground is not precisely leveled.

3. Results

3.1 Field design

Several designs of submerged ponds were successfully tested on the Abu Dhabi-type saltfish land. The design was modified to benefit from the tide and root movement rather than the use of pumps in the submerged ponds and was tested in Jubail, Saudi Arabia. The Sabkha lands are usually sedimentary, with a low filtration rate (less permeability) to the extent that 1 ha or more of a single irrigation outlet can be submerged, since the Sabkha field is not divided into separate basins, but the entire area is submerged.

In this method, the field is surrounded by a narrow barrier of soil and distributed within the barrier and at a depth of 1 m. A water pipe passes through the seawater during the tides to submerge the irrigation channels at a depth of half a meter to distribute the water to 10 m. A gate is also set up to control the level of water in the field. In order to fill the field, the gate is closed. The sea water (either as a result of the tide movement or using a pump) can pass through the main irrigation pipe to the irrigation channels and plant lines. A pump is often needed.

This method has been successfully tested in Mangrove, Salicornia, Abu Dhabi, where it has grown on Sabkha land, using seawater 50 g/l in irrigation. Soil salinity prior to planting was reduced from 80 to 120 g/l in a 10 cm layer by seawater washing for 1 week using 3 successive immersion and drainage cycles with salinity reduced to 50 g/l. The irrigation was done every 2–3 days in summer and every 4–5 days in winter.

It is worth mentioning that most of the water used is lost in surface drainage and therefore the water use efficiency is low. This design is suitable only in the case of Sabkha land, which can rely on the tidal and root water in irrigation without the need to use pumps.

It is the most widely used method of surface irrigation. This experiment was conducted on a sandy loam in Kino Bay, Mexico. Salicornia has been cultivated as an oilseed crop in 20 ha as a commercial farm since 1986. The farm was divided into

1 ha. The seawater well was used for irrigation from 5 to 10 days after germination and the water used was 3–4 m/200 days for yield. This rate falls within conventional crop rates, but irrigation efficiency is relatively low because about half of the added water is lost under the root zone.

3.2 Center pivot sprinkler irrigation

Salicornia has been cultivated, where watering can be used in the first 100 days (up to the floriculture stage) and then the pipes have been used to connect to connect the water to the ground level next to the growing plants. The amount of water used for irrigation and washing was about 2–3 m and the growth period of the crop is 250 days (about 1.25–1.5 times the evaporation rate). The machines and pipes used in this system must be resistant to the impact of seawater.

3.3 Drip irrigation

Drip irrigation using seawater is used to irrigate the Atriplex shrubs. A high yield was obtained. No salinity and clotting problems were observed. The irrigation is continuous daily, and salt accumulation is more frequent when burying the pores in the soil rather than on the soil surface.

Pibars and Mansour [30] compared sprinkler, surface drip, subsurface drip, and furrow irrigation to produce potato and sunflowers in the new reclaimed lands. Subsurface drip irrigation (SDI) with a 20-kPa irrigation criterion was among the most productive irrigation systems.

Pibars et al. [32] studied four options for managing drip irrigation of potatoes in North Dakota. Automation of the irrigation based on a soil water tension irrigation criterion at 30 kPa had relatively high water use efficiency. Tayel et al. [27] compared automated controlled SDI irrigation with the conventional semiclosed seepage subirrigation in Florida. The conventional irrigation system is under criticism because of surface runoff and nutrient contamination of adjoining waterways. The SDI system required more electrical energy but used 36% less water to obtain the same potato yield. Mansour et al. [42] examined irrigation-scheduling options for drip-irrigated potatoes. For sprinkler-irrigated potato, extensive work has been done on potato responses to N fertilizer and N losses, but relatively few studies have studied potato N fertilization and loss under drip irrigation.

Sprinkler irrigation at different irrigation criteria was compared to surface drip and buried drip irrigation (with a range of fertilization treatments), for potato yield and grade in Minnesota [3]. Less water was required using either drip irrigation system. Surface drip and buried drip were among the most productive systems for total and marketable yield. Furthermore, drip irrigation or sprinkler irrigation (at a relatively dry soil criteria) reduced nitrate leaching under potato compared to normal sprinkler irrigation [13] reporting that reduced nitrogen rates did not affect potato yield, when irrigated with a subsurface drip system. In **Figures 2–5**, the effect of different drip irrigation systems and different saline water on wheat grain, straw yield and wheat grain, and straw water productivity is shown.

Mansour and Aljughaiman [22] showed that drip irrigation had potential as an economically viable potato production method in the southeastern United States. Optimized irrigation rates were 99–86% of the water called for in their irrigation model. Ref. [29] examined tape depth and emitter spacing on tuber yield and grade of Norgold Russet potato in Lubbock, Texas. Tape depth or emitter spacing did not influence potato yield, but the proportion of misshaped tubers was greater when the tape was buried at 0.2 m than with shallower placement. Soil temperature was

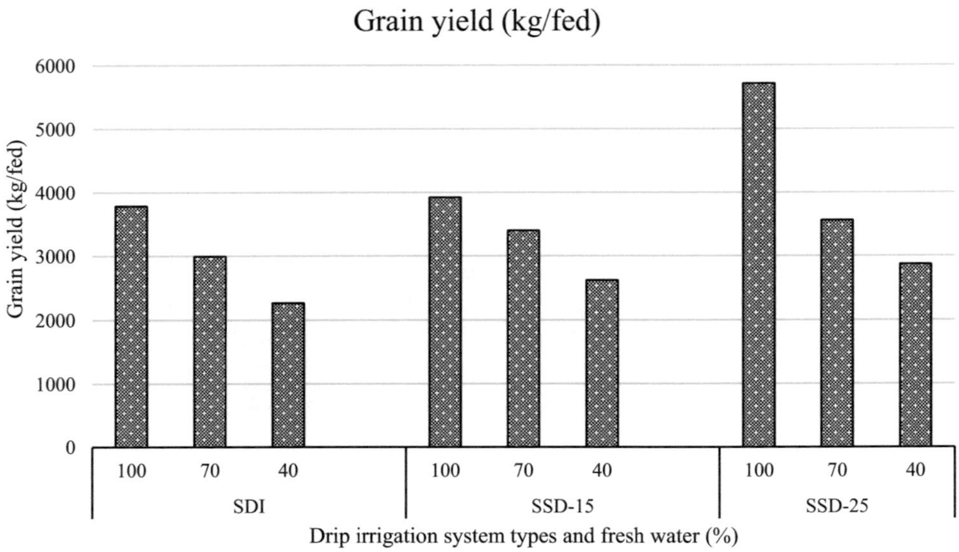

Figure 2.
Effect of different drip irrigation system types and different saline water on wheat grain yield. SDI: surface drip irrigation; SSD-15: subsurface drip irrigation at soil depth 15 cm; SSD-25: subsurface drip irrigation at soil depth 25 cm.

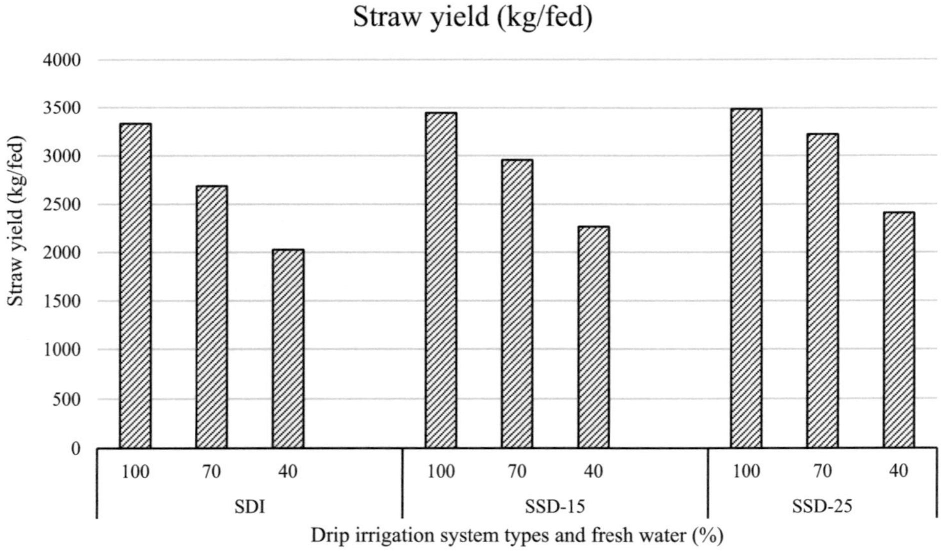

Figure 3.
Effect of different drip irrigation systems and different saline water on wheat straw yield. SDI: surface drip irrigation; SSD-15: subsurface drip irrigation at soil depth 15 cm; SSD-25: subsurface drip irrigation at soil depth 25 cm.

greater with the tape at 0.2 m than at 0.1 or 0.025 m. El-Hagarey et al. [28] found that tape depths of 0.08 m (above the seed piece) and 0.46 m (below the seed piece) performed better than intermediate and greater depths. Tayel et al. [38] used 10 drip irrigation treatments to examine the effect of the timing of irrigation deficits on potato yield and water use efficiency in Spain. Irrigation deficits occurring during mid- and late-season tuber bulking were particularly damaging to yield. High yield was combined with high water use efficiency when irrigation deficits were restricted early in the season. Mansour et al. [49] investigated the performance

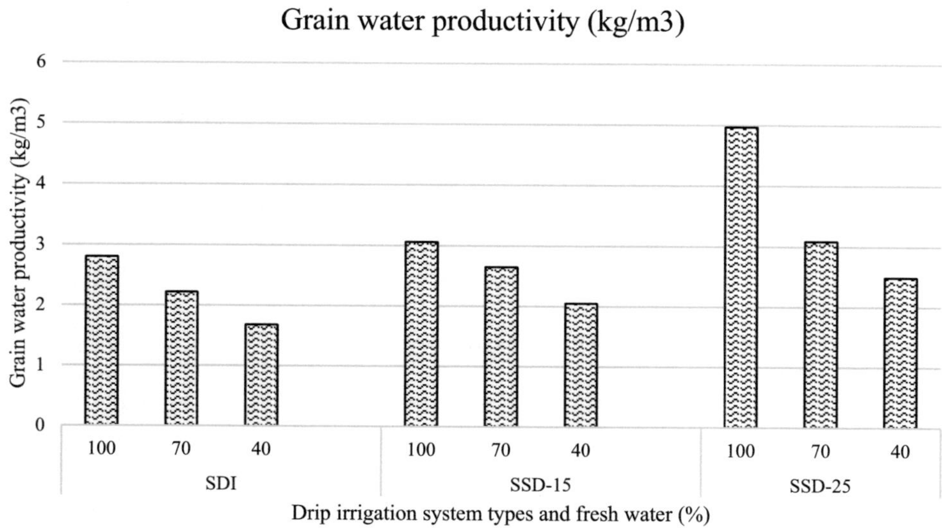

Figure 4.
Effect of different drip irrigation systems and different saline water on wheat grain yield water productivity. SDI: surface drip irrigation; SSD-15: subsurface drip irrigation at soil depth 15 cm; SSD-25: subsurface drip irrigation at soil depth 25 cm.

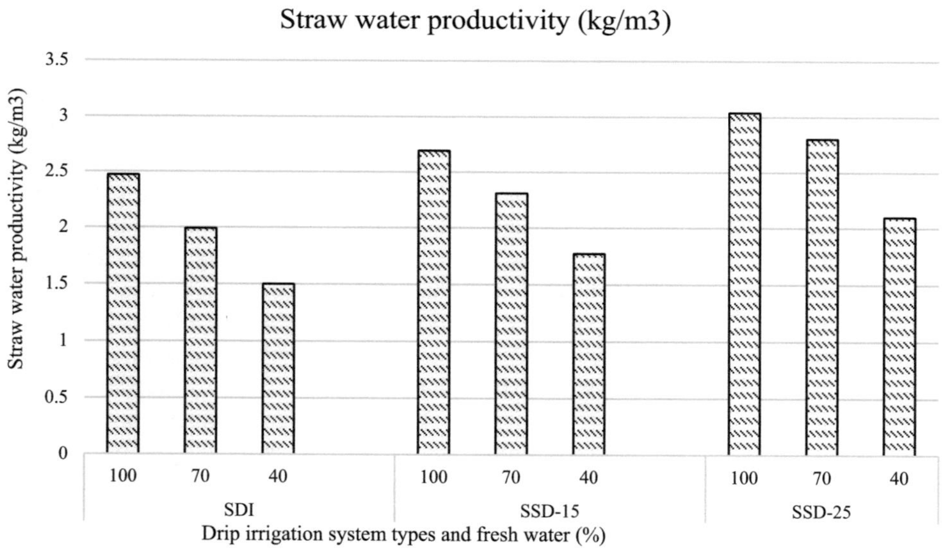

Figure 5.
Effect of different drip irrigation systems and different saline water on wheat straw yield water productivity. SDI: surface drip irrigation; SSD-15: subsurface drip irrigation at soil depth 15 cm; SSD-25: subsurface drip irrigation at soil depth 25 cm.

of 'Umatilla Russet' under drip irrigation in silt loam. The factors considered in the study were tape placement (one tape per row or one tape per two rows) and four soil water tension levels for automatically starting irrigation (1.5, 3.0, 4.5, and 6.0 bar). They concluded that drip tape placement had a significant effect on every variable except total marketable yield and bud-end fry color for which interactions of irrigation criteria with tape number were significant. Tape placement and irrigation criterion interacted to influence total yield, total marketable potatoes, and US No. 2 yield. Results indicated potato should be irrigated at 3.0 bar, given the silt loam soil and 2.5 mm water applied at each irrigation episode. The irrigation criterion considered alone only influenced the total US No. 1 and over 340 g tuber weight categories. Potato cultivars were very different in their performance under drip irrigation [54].

The accumulated experience shows that seawater irrigation depends on the efficiency of the irrigation system used, and as a result of the low efficiency of irrigation uses large amounts of seawater not to allow the depletion of water until it reaches the wilt point between the irrigation because in this case the concentration of salt will be very high in the region. The soil moisture should be kept close to the field capacity at any time. This means that the efficiency of each soil should be as high as possible.

3.4 Use of salt-loving plants as feed for animals

Halophyte plants are known to be a traditional source of animal feed, although some of the problems that accompany it include high concentrations of salts, low energy content, and low animal palatability compared with traditional fodder. For the cultivation of halophytes to be economically viable, their performance should be higher or at least equal to the traditional feed. Many studies have shown that, given the lack of adequate animal feeds, especially in desert conditions, certain varieties have been successfully cultivated and can be used as feed substitutes [54, 55].

It is important to keep in mind that if halophytes are used as feed, animals may need to increase the consumption of drinking water, and feed consumption per unit may increase in animal weight as a result of increased metal content in halophytes. The carcass fed to the ingredients of a diet containing halophytes is equal to that fed on traditional feeds. One of the most common halophytes studied and used is a Salicornia cultivar. The results of the University of Arizona on sheep showed that Salicornia (seeds and market) can be used, as well as the cut as an alternative to the processed barley or the cotton seed. Salicornia is cultivated to produce oil and straw seeds. Oil can be extracted from the seed age. Organic materials that are free of salts can be used in animal feed. Oil can also be used as a high energy source in animal feed, especially poultry. Mansour et al. [44] found that subsurface drip irrigation systems may increase water use efficiency due to reduced soil and plant surface evaporation and because only the root zone or the partial root zone is irrigated as opposed to sprinkler irrigation where the entire field area is wetted. Besides this physiological dimension, several studies have been conducted for development of irrigation systems for salinity management with drip irrigation using saline water [32]. According to [20], the DI permits a uniform and frequent application of water and a direct feeding of the plant at the root zone level, leading to an increase of yield and saving water [36]. According to [47], DI improves tomato yield and reduces leaf burn (browning). However, this system may result in localized accumulation of salts at the soil surface [72] due to increased evaporation. According to [39], salt accumulates on the soil surface before migrating and reaches the root zone when DI irrigation is used. Subsurface drip irrigation has been developed to improve salinity management and water use efficiency. According to [54], SDI decreases the accumulation of salts at the root zone level of plants, producing an improved yield and fruit quality.

4. Discussion

Electrolysis is mainly used to desalinate half-saline groundwater. Electrolysis usually occurs because the salt is dissolved in water, which decomposes into electrons (electrically charged particles) of sodium and chloride. Sodium ions carry a positive electrical charge and chloride ions carry a negative electrical charge. A wide chamber divided into several cells is used in electrolysis by means of thin plastic sheets called membranes. Two types of membranes are used, one of which allows only positive ions to pass through and the other passes only negative ions. There is a positive electrode in one of the terminal chambers and on the other end a negative electrode [19].

The freezing process depends on the established fact that ice crystals formed by salt water are saltfree. The most important disadvantages of this method are the problems caused by the transfer and purification of snow, and the most important advantages are reducing the deposition and corrosion as they are operating at relatively low temperatures [12–15].

The process of desalination is divided into two ways: direct freezing and indirect freezing in order to reduce costs and accelerate the provision of a larger volume of water needed, with the increasing need to exploit new sources of water to bridge the huge gap between supply and demand, to meet the need for development, and to meet population growth.

The matter of how to settle the desalination industry in the Arab world, especially in Egypt, was a target for the 11th Conference on Water Desalination, hosted by Cairo for the first time under the patronage of the Prime Minister [20–23]. About 400 Arab and international experts in water field and the organization of the Ministry of Housing represented by the Holding Company for Water and Drainage, in addition to the Saudi Arabia government, attended the event.

The conference, which was held under the slogan "Localization of the Desalination Industry in the Arab World", came under the attention of seawater as a source that can be exploited to obtain water through desalination processes to fill part of the water gap.

In recent years, the need for desalination technology in many Arab countries, notably Saudi Arabia and the Gulf States, has succeeded in developing policies and administrative and technical systems that have made a successful and distinctive experience. Other Arab countries have been stranded for various reasons, despite their maritime coasts, due to the difficulty of obtaining appropriate technology and high costs, and the absence of culture adequate appropriateness [24–26].

Mansour et al. [46], the conference's general secretary, stressed that water saving is one of the biggest challenges facing the Arab region, especially Egypt. It is suffering from a shortage of about 28 billion cubic meters annually, as well as rapid population growth, which requires additional new water resources, which led to the use of three new resources: desalination, reuse of wastewater, and groundwater. He added that it is necessary for all Arab countries to search for unconventional water resources under geographical determinants, including that most of the sources of rivers exist in non-Arab countries, as well as groundwater, as they are shared with other countries. He explained that the conference was prepared by a number of important international and Arab bodies, including Saudi Arabia, which produces about 1.60 billion cubic meters of water annually based on desalination technology and has considerable experience in this field, pointing out that the conference aims to benefit from these experiences for an institutional mechanism for cooperation with them and to promote the desalination industry in Egypt.

Considering that millions of cubic meters of freshwater are being wasted annually around the world, World Bank experts have alerted to the scarcity of water day after day, and experts fear that the world will not be able to provide the necessary water. He pointed out that by 2050, the availability of water will not exceed 10% of the available water a century ago (i.e., since 1950, most Arab countries are among the most arid regions of the earth, 15 countries are the poorest in water, and the Arab citizen will receive only 700 m^3 or about 80% of the water poverty limit, a thousand cubic meters per year). Zuhair stated that the desalination industry is one of the main industries in the Arab world, especially when there are 115 countries with desalination plants, but because the cost of this technology is expensive, this should stimulate us to continue scientific research to develop scientific solutions to reduce this cost.

Mansour et al. [40] state that if we take a closer look at a country with water shortages, such as Egypt, it will produce about 93 million cubic meters of desalinated water, which corresponds to the population of 93 million people in Egypt. One cubic meter per person per year, the steady increase in population far outweighs the increase in water production. "The settlement is still a controversial term," Fawzan said. "There are countries that have achieved a high degree of production, and other countries have gone a long way in scientific research," he said. "The resettlement of technology is still a dream that we are waiting for."

Abdul Majeed Al Awadhi, former Head of the Electricity and Water Authority in the Kingdom of Bahrain, talked about the desalination industry in the Gulf countries, pointing out that it reached about 70% of the water production in the United Arab Emirates (UAE), Qatar, and Bahrain and 40% in Saudi Arabia and Oman with an average of 60%. He explained that with the increasing demand for water, desalinated water is expected to become the main source in all Arab countries. He pointed out that the higher the desalination plants, the lower the cost, which currently ranges between one dollar and two dollars per cubic meter, and that there is a large gap between cost and revenue, which amounts to half a dollar per cubic meter. It is necessary to reduce this gap and provide a greater role for the private sector in this regard, according to his description. All desalination methods require large amounts of energy, and power generation is expensive whether it is generated by electric methods, by burning fuel, or by nuclear power plants.

Desalination may help mainly dry areas on the coasts but give little hope to overcome the scarcity of fresh water in cities that lie offshore or on mountains. Bringing water to these cities can be more expensive than desalination [15, 16].

The high cost of desalinating water is not important in places where only sea water is available. More than 200 water desalination plants have been established in the world, most notably in Saudi Arabia, Kuwait, Australia, California, Greenland, and some countries in South America. Some of these plants are small; many of which serve military centers in isolated places or serve as wells for desert diggers, island resorts, and industrial plants [17, 18].

The world's desalination plants produce more than 3.8 billion liters of fresh water per day. This production meets a fraction of the world's freshwater needs. A large water desalination plant, such as the one in Jubail, Saudi Arabia, has been designed to produce 950 million liters of fresh water per day. One of the most pressing dilemmas for humans is how to provide for the world's food and clothing needs, and the consequent provision of adequate natural resources, especially land and water, and thus provide adequate nutrition for the growing numbers of tropical and subtropical populations within the next 30 years [3].

5. Conclusion

FAO estimates the need for land resources at about 200 million hectares as new agricultural land to produce various crops. Only 93 million hectares of land can be used for agricultural expansion. Unfortunately, much of this space is now occupied by the forests that we must preserve to maintain the ecological balance and global climate of the entire planet. We can add to this difficult problem another factor the deterioration of fertile agricultural land, either as a result of destruction or salting or pollution in most of the lands of the countries of arid and semiarid regions, which stimulates the human need to find alternative sources of water and land to grow crops and increase vegetation.

In General, the Egyptian strategy should be not only focusing on water supply management but also managing water demand. Of which, 55 billion are from the Nile River and 5.2 from groundwater, the rest from rainwater and drainage, while desalinated water does not exceed 10% of water resources. The Egyptian water strategy should include increasing the amount of desalinated water to more than 50%, especially since Egypt is in a very rich location in saltwater sources that can be utilized to the maximum extent possible. Researchers have attempted to develop varieties of some traditional crops such as wheat that are saline resistant using local selective ecotourism techniques and using genetic engineering through which saline-tolerant genes are added, but it can be said that so far these efforts have not resulted in the production of candidate seawater breeds. The maximum salinity of irrigation water in the long term, even for the most salt-tolerant crops such as date palm, is still less than 5 mm. Sea salt salinity is between 35 and 40 mm. As we know, seawater is rich in sodium chloride, one of the most harmful substances to growing plants.

The use of nonconventional water resources and the preservation of what is already available are very important rules in all countries. One of the important ideas of the last half of the twentieth century was the use of seawater, the first serious appearance of the idea after the Second World War. The use of seawater is very possible in sandy and desert environments. The sea aids in the development of saline-tolerant crops in irrigated land. This idea is an ideal solution, with 97% of the world's water being brackish water (seas and oceans), while desert lands are also widespread, accounting for 43% of the land area.

Acknowledgements

The authors would like to thank the "Grassland Talents" project, Inner Mongolia (2016) 40, Xilingol Vocational College; Talented Young Scientist Program (Egypt-18-050), CHINA; and the National Research Centre, EGYPT, for the support and fund for the publishing process of this research work.

Author details

Hani Abdelghani Mansour[1,2*], Ren Hongjouan[2], Hu Jiandong[2], Bao Hong Feng[2] and Liang Changmei[2]

1 Department of Water Relations and Field Irrigation, Agriculture and Biological Research Division, National Research Centre, Cairo, Egypt

2 Xilingol Vocational College, Xilinhot City, Inner Mongolia, China

*Address all correspondence to: ha.mansour@nrc.sci.eg

References

[1] American Society of Agricultural Engineering. Agricultural Engineering, Year Book; ASAE D 230.2. St. Joseph, Michigan: ASAE; 1975

[2] DeTar WR, Browne GT, Phene CJ, Sanden BL. Real-time irrigation scheduling of corn with sprinkler and subsurface drip systems. In: Camp CR, Sadler EJ, Yoder RE, editors. Proc Int'l Conf. on Evapotranspiration and Irrigation Scheduling. St. Joseph, Michigan: ASAE; 1996. pp. 812-824

[3] Dhuyvetter KC, Lamm FR, Rogers DH. Subsurface drip irrigation (sdi) for field corn—An economic analysis. In: Proceedings of the 5th International Micro-Irrigation Congress; 2-6 April 1995; Orlando, Florida. 1995. pp. 395-401

[4] Doorenbos J, Pruitt WO. Crop Water Requirements Irrigation and Drainage. Rome, Italy: Food and Agricultural Organization (FAO) of the United Nations; 1977. p. 24

[5] El Amami H, Zairi A, Pereira LS, Machado T, Slatni A, Rodrigues P. Deficit irrigation of cereals and horticultural crops: Economic analysis. Agricultural Engineering International: The CIGR Journal of Scientific Research and Development. 2001;**3**(1):1-11. Manuscript LW 00 007b

[6] Fabeiro C, Martin de Santa Olalla F, de Juan JA. Yield and size of deficit irrigated corn. Agricultural Water Management. 2001;**48**:255-266

[7] Gee GW, Bauder JW. Particle size analysis. In: Klute R, editor. Methods of Soil Analysis. Part 1 (Agron). 2nd ed. Madison, WI: ASA and SSSA; 1986. pp. 383-412

[8] Mansour HA, Saad A, Ibrahim AAA, El-Hagarey ME. Management of Irrigation System: Quality Performance of Egyptian Wheat. Micro Irrigation Management: Technological Advances and their Applications. 2016. p. 279.2

[9] Mansour HA, Abd El-Hady M, Bralts VF, Engel BA. Performance automation controller of drip irrigation systems using saline water for wheat yield and water productivity in Egypt. Journal of Irrigation and Drainage Engineering. 2016;**142**(10):1-6

[10] Mansour HA, Pibars SK, Gaballah MS, Mohammed KAS. Effect of different nitrogen fertilizer levels, and wheat cultivars on yield and its components under sprinkler irrigation system management in sandy soil. International Journal of ChemTech Research. 2016;**9**(9):1

[11] Mansour HAA, Mehanna HM, El-Hagarey ME, Hassan AS. Automation of mini-sprinkler and drip irrigation systems. In: Closed Circuit Trickle Irrigation Design: Theory and Applications. Florida, USA: Apple Academic Press, Taylor and Frances; 2015. p. 179

[12] Mansour HA, Abdallah EF, Gaballah MS, Gyuricza C. Impact of bubbler discharge and irrigation water quantity on 1-hydraulic performance evaluation and maize biomass yield. International Journal of GEOMATE. 2015;**9**(2):1538

[13] Mansour HA, Abdel-Hady M, Eldardiry EI, Bralts VF. Performance of automatic control different localized irrigation systems and lateral lengths for: 1-Emitters clogging and maize (*Zea mays* L.) growth and yield. International Journal of GEOMATE. 2015;**9**(2):1545

[14] Mansour HAA, Tayel MY, Lightfoot DA, El-Gindy AM. Energy and water savings in drip irrigation systems. In:

Closed Circuit Trickle Irrigation Design: Theory and Applications. Florida, USA: Apple Academic Press, Taylor and Frances; 2015. p. 149

[15] Mansour HAA. Design considerations for closed circuit design of drip irrigation system. In: Closed Circuit Trickle Irrigation Design: Theory and Applications. Florida, USA: Apple Academic Press, Taylor and Frances; 2015. p. 61

[16] Mansour HAA, El-Hady MA, Gyurciza CS. Water and fertilizer use efficiencies for drip irrigated maize. In: Closed Circuit Trickle Irrigation Design: Theory and Applications. Florida, USA: Apple Academic Press, Taylor and Frances; 2015. p. 207

[17] Mansour HA, Pibars SK, Bralts VF. The hydraulic evaluation of MTI and DIS as a localized irrigation systems and treated agricultural wastewater for potato growth and water productivity. International Journal of ChemTech Research. 2015;**8**(12):142

[18] Mansour HAA, Aljughaiman AS. Water and fertilizer use efficiencies for drip irrigated corn: Kingdom of Saudi Arabia. In: Closed Circuit Trickle Irrigation Design: Theory and Applications. Florida, USA: Apple Academic Press, Taylor and Frances; 2015. p. 233

[19] Mansour HA. Performance automatic sprinkler irrigation management for production and quality of different Egyptian wheat varities. International Journal of ChemTech Research. 2015;**8**(12):226

[20] Mansour HAA, El-Melhem Y. Performance of drip irrigated yellow corn: Kingdom of Saudi Arabia. In: Closed Circuit Trickle Irrigation Design: Theory and Applications. Florida, USA: Apple Academic Press, Taylor and Frances; 2015. p. 219

[21] Mansour HA, Pibars SK, Abd El-Hady M, Eldardiry EI. Effect of water management by drip irrigation automation controller system on faba bean production under water deficit. International Journal of GEOMATE. 2014;**7**(2):1047

[22] Mansour HA, Aljughaiman AS. Water and fertilizers use efficiency of corn crop under closed circuits of drip irrigation system. Journal of Applied Sciences Research. 2012;**8**(11):5485-5493

[23] Tayel MY, Shaaban SM, Eldardiry EA, Mansour HA. Wheat yield versus seed bed conditions. Bioscience Research. 2018;**15**(3):1943-1951

[24] Ibrahim A, Csúr-Varga A, Jolánkai M, Mansour H, Hamed A. Monitoring some quality attributes of different wheat varieties by infrared technology. Agricultural Engineering International: CIGR Journal. 2018;**20**(1):201

[25] Tayel MY, Shaaban SM, Mansour HA, Abdallah EF. Response of fodder beet plants grown in a sandy soil to different plowing conditions. International Journal of ChemTech Research. 2016;**9**(9):20

[26] Pibars SK, Mansour HA. Evaluation of response sesame water productivity to modern chemigation systems in new reclaimed lands. International Journal of ChemTech Research. 2016;**9**(9):10

[27] Tayel MY, Mansour HAA, Pibars SK. Performance of drip irrigated soybean. In: Closed Circuit Trickle Irrigation Design: Theory and Applications. Florida, USA: Apple Academic Press, Taylor and Frances; 2015. p. 250

[28] El-Hagarey ME, Mehanna HM, Mansour HA. Soil moisture and salinity distributions under modified sprinkler irrigation. In: Closed Circuit Trickle Irrigation Design: Theory and

Applications. Florida, USA: Apple Academic Press, Taylor and Frances; 2015. p. 3

[29] Tayel MY, Mansour HAA, Pibars SK. Performance of sprinkler irrigated wheat—Part I. In: Closed Circuit Trickle Irrigation Design: Theory and Applications. Florida, USA: Apple Academic Press, Taylor and Frances; 2015. p. 23

[30] Pibars SK, Mansour HA. Evaluate the response of sunflower productivity to modern chemigation systems in new reclaimed lands. International Journal of ChemTech Research. 2015;**8**(12):160

[31] Tayel MY, Pibars SK, Mansour HAA. Evaluation of emitter clogging. In: Closed Circuit Trickle Irrigation Design: Theory and Applications. Florida, USA: Apple Academic Press, Taylor and Frances; 2015. p. 287

[32] Pibars SK, Mansour HAA, Abd El-Hady M, Eldardiry EEI. Evaluation of emitter clogging for drip irrigated snap beans. In: Closed Circuit Trickle Irrigation Design: Theory and Applications. Florida, USA: Apple Academic Press, Taylor and Frances; 2015. p. 273

[33] Tayel MY, Mansour HAA, Pibars SK. Performance of maize under bubbler irrigation system. In: Closed Circuit Trickle Irrigation Design: Theory and Applications. 2015. p. 135

[34] Goyal MR, Mansour HAA. Closed Circuit Trickle Irrigation Design: Theory and Applications. Florida, USA: Apple Academic Press, Taylor and Frances; 2015. p. 1

[35] Eldardiry EE, Hellal F, Mansour HAA. Performance of sprinkler irrigated wheat—Part II. In: Closed Circuit Trickle Irrigation Design: Theory and Applications. Florida, USA: Apple Academic Press, Taylor and Frances; 2015. p. 41

[36] Tayel MY, Shaaban SM, Mansour HA. Effect of plowing conditions on the tractor wheel slippage and fuel consumption in sandy soil. International Journal of ChemTech Research. 2015;**8**(12):151

[37] Tayel MY, El-Gindy AM, Mansour HA. Effect of drip irrigation circuits design and lateral line length on: IV-uniformity coefficient and coefficient of variation. Journal of Applied Sciences Research. 2012;**8**(5):2741

[38] Tayel MY, El-Gindy AM, Mansour HA. Effect of drip irrigation circuits design and lateral line length on: III-dripper and lateral discharge. Journal of Applied Sciences Research. 2012;**8**(5):2725

[39] Mansour HA, Pibars SK, Gaballah MS. Effect of the smart drip irrigation system and water deficit on contour maps of soil moisture distribution. World Wide Journal of Multidisciplinary Research and Development. 2018;**4**(2):313-320

[40] Mansour HA, ELMesiry TA, Abdelhady AA. The effect of operating speed and olive varieties harvested by hand-held machine on productivity and damage percentage. World Wide Journal of Multidisciplinary Research and Development. 2018;**4**(2):355-360

[41] Mansour HA, Elmesiry TA, Abdelhady AA. Cost analysis of olive harvesting by hand-held machine. World Wide Journal of Multidisciplinary Research and Development. 2018;**4**(2):361-363

[42] Mansour HA, Eldardiry EI, Abd El-Hady M. Validation by aqua crop model for Faba bean yield, water productivity under smart controller of drip irrigation system and compost fertilizer in sandy soil. European Journal of Academic Essays. 2018;**5**(5):138-144

[43] Mansour HA, Camelia YE-D, Mohammed KAS, Selim E-MM. Response of corn plants to modified closed circuits of bubbler irrigation system and nitrogen fertilization rates. European Journal of Academic Essays. 2018;**5**(5):150-158

[44] Mansour HA, Eldardiry EI, Abd El-Hady M. The effect of different irrigation systems, nitrogen levels and Egyptian wheat varieties on flour quality parameters. European Journal of Academic Essays. 2018;**5**(5):159-167

[45] Mansour HA, Eldardiry EI, Abd El-Hady M. Response of Egyptian wheat varieties to irrigation systems and nitrogen fertilization rates. European Journal of Academic Essays. 2018;**5**(5):168-176

[46] Mansour HA, Abd-Elmaboud SK. Effect of sub-surface drip irrigation on moisture and salinity distribution under different water deficit and depths. European Journal of Academic Essays. 2018;**5**(5):177-185

[47] Mansour HA, Abdel Aziz AA, Tayel MY, El-Gindy AM. Effect of Localized Irrigation Systems on Productivity and Quality of Grape Crop. Research Bulletin. Faculty of Agriculture, Ain Shams University; 2005

[48] Mansour HA. The response of grape fruits to application of water and fertilizers under different localized irrigation systems [M.Sc. thesis]. Egypt: Faculty of Agriculture, Ain Shams University; 2006

[49] Mansour HA, Tayel MY, El-Hady MAA, Lightfoot DA, El-Gindy AM. Modification of water application uniformity among closed circuit trickle irrigation systems. Agricultural Science Journal. 2010;**1**(1):1-9

[50] Mansour HA, Tayel MY, Lightfoot DA, El-Gindy AM. Energy and water saving by using closed circuits of drip irrigation systems. Agricultural Science Journal. 2010;**1**(3):154-177

[51] Mansour HA. Design considerations for closed circuits of drip irrigation system [PhD thesis]. Egypt: Faculty of Agriculture, Ain Shams University; 2012

[52] Mansour HA, Mehanna HM, El-Hagarey ME, Hassan AS. Using automation controller system and simulation program for testing closed circuits of mini-sprinkler irrigation system. Open Journal of Modelling and Simulation. 2013;**1**(2):14-23

[53] Mansour HA. Evaluation of closed circuits drip irrigation by using simulation program under automation controller. International Journal of Control, Automation and Systems. 2013;**2**(3):128-136

[54] Mansour HA, Gyuricza C. Impact of water quality and automation controller drip irrigation systems on potato growth and yield. International Journal of Advanced Research. 2013;**1**(9):235-244

[55] Mansour HA, Abd El-Hady M, Gyurciza CS. Effect of localized irrigation systems and humic compost fertilizer on water and fertilizer use efficiency of maize in sandy soil. Australian Journal of Agricultural Research. 2013;**2**(10):292-297

[56] Mansour HA, El-Melhem Y. Impact the automatic control of closed circuits drip irrigation system on yellow corn growth and yield. International Journal of Advanced Research. 2013;**1**(10):33-42

[57] Pibars SK, Mansour HA, Gaballah MS. Effect of water regime and compost tea on growth characters of sugar beet under different irrigation system. World Wide Journal of Multidisciplinary Research and Development. 2018;**4**(2):321-325

[58] Abd El-Hady M, Mansour HA, Abdelghany AE. Improvement of different crop seeds germination through by magnetized saline water treated. World Wide Journal of Multidisciplinary Research and Development. 2018;**4**(1):57-63

[59] Abd El-Hady M, Mansour HA, Abdelghany AE. The effect of electromagnetic forces and saline water on hydraulic properties of drip irrigation system and soil salinity distribution. World Wide Journal of Multidisciplinary Research and Development. 2018;**4**(1):79-88

[60] El-Gindy AM, Arafa YE, Abd El-Hady M, Mansour HA, Abdelghany AE. Effect of drip irrigation system salinity and magnetic water treatment on turnip yield and yield characters. World Wide Journal of Multidisciplinary Research and Development. 2018;**4**(1):89-96

[61] Eldardiry E, Hellal F, Mansour H, El Hady MA. Assessment cultivated period and farm yard manure addition on some soil properties, nutrient content and wheat yield under sprinkler irrigation system. Agricultural Sciences. 2013;**4**(1):14-22

[62] El-Hagarey ME, Mehanna HM, Mansour HA. Movable surface irrigation system (MSIS) impact on spatial and temporal distribution of soil moisture and salinity. IOSR Journal of Agriculture and Veterinary Science (IOSR-JAVS). 2016;**7**(6):49-57

[63] El-Hagarey ME, El-Nesr MN, Mehanna HM, Mansour HA. Energy, economic analysis and efficiencies of micro drip irrigation II. Economic analysis and efficiencies. IOSR Journal of Agriculture and Veterinary Science (IOSR-JAVS). 2016;**7**(8):19-26

[64] El-Hagarey ME, El-Nesr MN, Mehanna HM, Mansour HA. Energy, economic analysis and efficiencies of micro drip irrigation I. Energy analysis. IOSR Journal of Agriculture and Veterinary Science (IOSR-JAVS). 2016;**7**(8):11-18

[65] Tayel MY, Abdel-Ghani Mansour H, Pibars SK. Effect of closed circuits drip irrigation system and lateral lines length on growth, yield, quality and water use efficiency of soybean crop. Agricultural Sciences. 2013;**4**(2):85-90

[66] Tayel MY, Pibars SK, Mansour HA-G. Effect of drip irrigation method, nitrogin source, and flushing schedule on emitter clogging. Agricultural Sciences. 2013;**4**(3):131-137

[67] Sabreen K, Mansour HA, Abdel Hady M, Eldardiry EE. Maximize utilization from fertigation management for snap bean (*Phaseolus vulgaris* L.) under sandy soil. IOSR Journal of Agriculture and veterinary Science (IOSR-JAVS). 2016;**7**(7):25-32

[68] Tayel MY, Mansour HA. Effect of drip irrigation circuit design and lateral line length on V. Water and fertilizer use efficiency. Current Advances in Environmental Science. 2013;**1**(1):1-6

[69] Tayel MY, El-Gindy AA, Abd El-Hady M, Mansour HA. Effect of irrigation systems on: III. Some hydrophysical efficiency as affected by irrigation methods. American-Eurasian Journal of Agricultural & Environmental Sciences. 2006;**1**:276-285

[70] Tayel MY, El-Gindy AA, Abd El-Hady M, Mansour HA. Effect of irrigation systems on: I. Yield, water and fertilizer use efficiency of grape. Journal of Applied Sciences Research. 2007;**3**(5):367-372

[71] Tayel MY, Mansour HA, Lightfoot DA. Effect of drip irrigation circuit design and lateral line lengths i- on pressure head and friction loss. Agricultural Sciences. 2012;**3**(3):392-399

[72] Tayel MY, Mansour HA, Lightfoot DA. Effect of drip irrigation circuit design and lateral line lengths ii- on flow velocity and velocity head. Agricultural Sciences. 2012;**3**(4):531-537

[73] Tayel MY, Sabreen KP, Mansour HA. The effect of irrigation intervals and tillage systems on cotton crop: 1-Vegetative growth. Agricultural Water Management. 2013;**1**(9):286-295

[74] Tayel MY, Mansour HA, Pibars SK. Impact of bubbler irrigation system discharges and irrigation water amounts on maize: 1-Irrigation efficiency, growth, grain and biomass production. International Journal of Advanced Research. 2016;**2**(1):716-724

[75] Tayel MY, Mansour HA, Pibars SK. Effect of two sprinkler irrigation types on wheat i- uniformity, vegetative growth and yield. International Journal of Advanced Research. 2016;**2**(1):47-56

[76] Tayel MY, Sabreen KP, Mansour HA. Response of cotton crop to irrigation intervals and tillage treatment: 2. Yield and water use efficiency. International Journal of Advanced Research. 2016;**2**(1):725-732

Chapter 5

Vulnerability of Environmental Resources in Indus Basin after the Development of Irrigation System

Muhammad Irfan, Abdul Qadir, Habib Ali, Nadia Jamil and Sajid Rashid Ahmad

Abstract

The climatic and topographic characteristics of Indus Basin provided an excellent condition for the development of irrigation system. Archaeological remains of Harappa and Mohenjo-Daro indicated that several canals were constructed in this region. The Indus River System (IRS) was developed into a complex network of canals, and 74% of its water was utilized for irrigation after Indus Water Treaty. After 1947, Indus irrigation network was extended, and cropland area was increased from 8.5 to 18.2 MH in Pakistan and 2.02 to 8.5 MH in India. Construction of dams, barrages, and canals to divert the maximum river water for irrigation resulted in drying up the natural pathways of the rivers, except during monsoon season. The aquifer in the irrigated areas became high and created problems of waterlogging and salinity, but due to extensive groundwater extraction, water table near urban centers is lowered now. Water quality was degraded due to addition of fertilizers, pesticides, chemicals, municipal sewage, and industrial effluents. Due to climate change, the glaciers in the upper catchment areas are continuously retreating and the frequency of floods and droughts is increasing. The objective of this chapter is to provide a comprehensive review of irrigation system developments in Indus Basin and its implications on environmental resources.

Keywords: Indus River, trans-boundary rivers, cultivation, ecological degradation

1. Introduction

About 50 million years ago, in the Mesozoic era, the shallow sandy Tethys Sea upfolded and formed the Great Himalayan Ranges because of the collision of Indian plate and the Siberian plate. The Indus basin comprised of lofty Himalayan mountains in the north and flat plains of Punjab and Sindh in the east and south. These mountains with immense snow cover gave birth to the Indus River and its tributaries [1]. The Indus River originates from Lake *Manasarovar* in Tibet, China, which traverses a total length of 3200 km (**Figure 1**). From the point of origin, the river flows in the northwest direction and then turns southward after reaching the Hindu Kush Mountains [2]. Many smaller tributaries join the Indus River on its way including *Shyok*, *Zanskar*, *Gilgit*, *Swat*, and *Kabul*. Near *Kalabagh*, it enters into the alluvial plain of Punjab. Five tributaries, viz., Jhelum, Chenab, Ravi, Beas, and Sutlej, join and traverse in the form of Panjnad and then join the Indus River. The mighty river runs as a large single channel through the Sindh province and ends up draining into the Arabian Sea [3]. The alluvial plain of Indus River is further divided into upper and lower Indus basins based on topography and elevation. The upper Indus basin comprises of the high-altitude mountainous areas with rugged topography including western areas of Tibet and Ladakh, stretching to the Himalayan foothills. The lower Indus basin starts from *Attok* and ends at Indus delta and consists of vast plains of Punjab and Sindh. The climate of Indus basin is humid in the north, semiarid in Punjab, and arid in the Sindh province [4]. The Indus River shares 52% of water in Indus River system, whereas the rest of the 48% is contributed by the Indus River tributaries. The rivers of the Indus basin receive more than 50% water from the glaciers followed by well-defined monsoon system in the upper catchment during monsoon season. The data of the past one century highlights slight reduction in the water flow in the Indus River system (**Table 1**).

The regular flow in Indus basin rivers provided the conducive conditions for early human settlements. One of the oldest civilizations, Indus valley civilization emerged in western South Asia in the prehistoric era along the Indus River and its tributaries when fertile land converted into cropland to grow wheat and barley and pasture to rear the cattle, sheep, and goats [15]. Since that time, several cultures, languages, and religions have emerged, invaded, or mixed because of the good climatic conditions [16]. Several invasions have been documented in history since the prehistoric times, and invaders settled to use the natural and water resources for prospering life. The Indus River basin favored the development of a large irrigation system. In the British era, the irrigation system was developed to increase the crop production in order to develop the agriculture-based economy, which turned the basin into a densely populated area. By the development of irrigation system and introduction of fertilizers and pesticides, agricultural production increased many

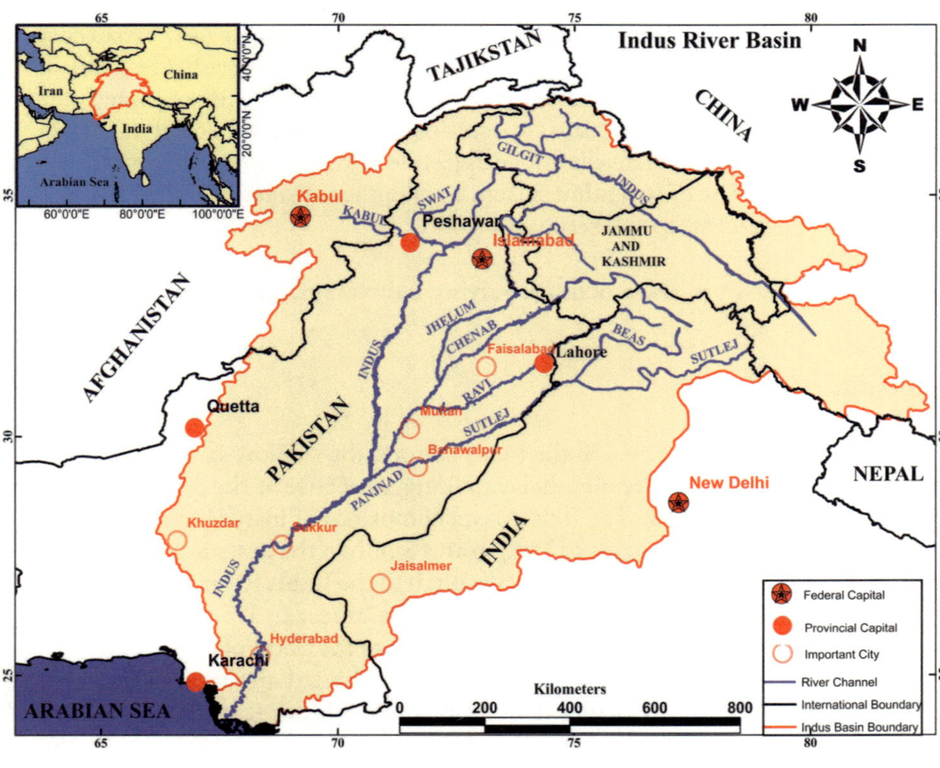

Figure 1.
Map representing river system, major cities, and the extent of Indus River basin.

River	Water Share in IRS	Share from snow and glaciers	Annual flow During monsoon season	Mean Annual Flow				Catchment Area (km²)	Agricultural Area (Million hectares)	
				Before IWT (1922-1960)		After IWT (1961-2010)				
				MAF*	BCM**	MAF	BCM		1947	2015
Indus	52%	65%	54%	90	111	88	108	466,000	In Pakistan	
Jhelum	16%	50%	36%	23	28	22	27	63,500	8.5	18.2
Chenab	13%	49%	56%	25	31	25	31	67,600		
Sutlej	8%	55%	62%	14	17	13	16	121,988	In India	
Ravi	4%	-	51%	6	8	7⁺	8.5⁺	40,400	2.02	8.5
Beas	7%	35%	67%	13	16	12	14.8	20,303		

*MAF= Million Acre Feet **BCM=Billion Cubic Meter ⁺Discharge from Ranjit Sagar Dam at gauging station near Mukteshwar

Table 1.
Water division, glacial contribution, mean annual flow, catchment area, and agricultural area of Indus River system (IRS) [5–14].

folds. This extensive human intervention in the Indus basin resulted in the adverse effects on the ecosystem of the Indus plain from the Himalayas to Indus delta [3].

Mohenjo-daro and *Harappa* are believed to be the two main centers of the Indus valley civilization. The ruins of *Harappa* excavation in the Punjab province highlighted that the grain commodity trade evolved which led to the development of earlier urban centers. The second center, *Mohenjo-daro*, found in the Sindh province, Pakistan, suffered from the change in the path of the Indus River. It is also believed that the continuous salinity problems and heavy flooding resulted in the collapse of Indus valley civilization [17, 18]. Early Harappans developed hydrological engineering practices like constructing *gabarbands* and dry masonry dams to divert the water toward the agricultural fields and digging wells and *karez* for the proper utilization of ground and river water. *Karez* system is still in use in Balochistan, while *gabarbands* are widely distributed in Balochistan and Kohistan [19].

2. Historical developments in the Indus basin irrigation system

In the development of early irrigation canals, the rulers of states in upper Indus basin areas played a vital role and paved the way to the construction of the complex irrigation system that exists today. Areas where significant work was done include Jammu and Kashmir regions, Punjab, and Sindh. Later on, further developmental work was carried out in several stages and different eras. Even today, an expansion of Indus irrigation system could be observed in different regions of Pakistan and India.

2.1 Early developments in Jammu and Kashmir

Development of the canal irrigation system started in the eighth century in the Kashmir region after the regular flooding of the valley and rise in the prices of crops that made the survival of the poor very difficult. In such a situation, King *Lalitaditya* set the drainage system in a suitable direction and distributed the Jhelum River water to different villages for increased production of crops. The Martand Canal was constructed in that period; starting from the left bank of *Liddar* River and terminating at *Karewa* of *Martands*, the canal irrigates about 3844.5 ha of land along its 50 km long channel. King *Avantivarman* started the channel cleaning, broadening, construction of stone embankments, and changing of the location of river junction. Villages were protected by constructing dykes, and in order to make the irrigation system of the valley more effective, a network of canals was constructed, and rain-dependent areas were provided with irrigation facilities. The *Nur* and *Nandikul* Canals were constructed with a length of 13 and 30 km, respectively. *Nur* canal originated from the Jhelum River and terminated at *Anderkot* Village, while *Nandikul* Canal was designed to irrigate 3237.5 ha of land of the northern *Anantnag* by receiving its water from the Nullah Anantnag. Sultan *Zain-ul-Abidin* constructed the *Zainagir* Canal on the *Madhumati* River, to irrigate the rice fields and apple orchards on 5463.3 ha in the district *Baramulla* [20].

2.2 Early irrigation system in Punjab

The Punjab province is located downstream to the Kashmir, and early developments in irrigation system of this region occurred during the thirteenth to sixteenth century. The first canal in Punjab was constructed about five centuries ago by the Mughal Emperors. In the beginning, only inundation channels were designed to deliver river water to the cropland during the high flow season. Because of some technical problems, viz., unpredicted high flow season, siltation, and breaching, these inundation canals could not deliver the water effectively to the cropland. Later on, some primitive types of headworks on rivers were constructed to get more control over the water. These headworks either did not extend across the entire stream or allowed the floods to pass over their crests. Mughal Emperor Jahangir constructed an 80 km perennial canal on Ravi River to deliver the water to the Gardens of Sheikhupura [21]. In 1643, Shah Nahr developed to provide water to the Shalamar gardens and other irrigational lands around Lahore city [22]. Later on, *Hajiwah* and *Tiwana* Canals were developed in Punjab with support of British rulers to irrigate croplands in the *Mailsi* and *Sargodha* regions. Furthermore, 11 canals were built on the Indus River to irrigate the agricultural lands of D.G. Khan Region during 1875 [23].

2.3 The modern irrigation system in Punjab and Sindh

The development of the existing modern Indus irrigation system started in the mid of nineteenth century during the British rule. Food demand and British economic interests in the agricultural products specifically cotton were a major driving force for the development of an extensive agriculture system in British India. A large number of inundation canals originating from the Indus River system were remodeled [24]. To ensure the water supply in cropland of Punjab, Sindh, and Khyber Pakhtunkhwa (KPK), several permanent headworks were constructed. The construction of Marala Headworks started on Chenab River in 1887 to irrigate the Upper Rachna Doab through Upper Chenab Canal, whereas, in 1890, Chenab water diverted to Sandal Bar from Lower Chenab Canal. In 1897, Rasul Headworks were constructed on the Jhelum River to feed the Lower Jhelum Canal to irrigate the agricultural land in the Chaj Doab [21]. In 1902, Madhopur Headworks were completed on the Ravi River to provide water for agriculture to the Upper Bari Doab. Furthermore, the Triple Canal Project was designed and sanctioned in 1905. It was the first project to transfer the river water from one to another river. A gate-regulated canal, the Upper Jhelum Canal, was designed to provide irrigation water to almost 139,212 ha per annum on its way from Mangla to Khanki. The construction work of the canal was completed in 1917 and drained its water into Chenab River in the upstream of the Khanki Barrage of Lower Chenab Canal. The second canal was the Upper Chenab Canal, originating 58 km upstream of Khanki from Marala Barrage, designed to irrigate almost 262,236.7 ha of cropland on its way from Marala to Balloki. The canal opened for irrigational purposes in 1912, and its water drained into the Ravi River above Balloki Headworks. The third canal, the Lower Bari Doab canal, originated from a 0.5 km long weir on the Ravi River near Balloki and irrigated almost 354,910 ha of lands in Montgomery District (Sahiwal) and Multan. It was one of the major irrigational projects executed during the British Era [24].

After the First World War, the British Government decided to remodel the pre-existing inundation canals along with the construction of new canals to irrigate the parched areas of Sutlej Valley in 1921. The project was aimed to provide perennial water supply to the inundation canals after remodeling them by controlling the river water with the help of barrages along with irrigating more and more areas of land. Construction of four barrages along with 11 canals on the Sutlej River was completed in 1933 to convert the arid land into cropland. The Ferozepur Barrage with three non-perennial canals, namely, Bikaner Canal, Eastern Canal, and Dipalpur Canal, was constructed near Ferozepur to irrigate cultivated areas of Bikaner State, Ferozepur district, northeastern areas of the Bahawalpur State, and Lahore and Montgomery (Sahiwal) Districts. Sulemanki Barrage with three perennial canals, namely, Eastern Sidiqia Canal, Fordwah Canal, and Pakpattan Canal, was constructed to irrigate some areas of the Bahawalpur State. Furthermore, Islam Barrage was constructed in Tehsil Hasilpur with three non-perennial canals, namely, Mailsi Canal, Qaimpur Canal, and Bahawal Canal, irrigating about 577,892 ha of cropland. After the confluence of Sutlej River and Chenab, Panjnad Barrage was constructed with a perennial canal (Abbasia Canal) and a non-perennial canal (Panjnad Canal) to irrigate 44,920 ha and 541,875 ha of cropland, respectively [24]. In 1922, Maharaja Ganga Singh constructed a canal to irrigate the Bikaner State originating from the left bank of the Sutlej River [25].

The Sukkur Barrage Project was the first-ever barrage to be built on the Indus River sanctioned in 1923 and completed in 1932 with seven canals [26]. The Trimmu Barrage with three canals was constructed on the Chenab River below the confluence of the Jhelum River during 1937–1939. It was the last barrage completed before the start of the Second World War. During the partition of subcontinent, the construction of Jinnah Barrage and Kotri Barrage on the Indus River was in progress. At that time, the Bhakra Dam was also under construction on the Sutlej River. Jinnah Barrage was completed in 1947, and Kotri Barrage was completed in 1955 [21].

2.4 Post-partition developments

Due to the partition of India and Pakistan, Ferozepur and Madhupur Headworks became the part of India, which triggered the Indus water dispute. India cut off the water supplies of Upper Bari Doab canal and made all of the downstream irrigation activities impossible to be carried out. In this situation of water scarcity, Pakistan immediately constructed Bombanwali-Ravi-Bedian (BRB) Link Canal to provide water supply to irrigate the Upper Bari Doab. It was a 164 km canal originating from Upper Chenab Canal and moving southward to Bedian. India also constructed two main canals from Sutlej River in order to divert the water of the river from flowing downstream into Pakistan. To maintain the water level in Sutlej River, the Balloki-Sulemanki Link canal was constructed from Balloki Headworks on Ravi River to Sulemanki Headworks on the Sutlej River. Both the BRB canal and Balloki-Sulemanki canals were completed within a duration of 3 years (1951–1954). Before the war of 1965, another canal named Marala-Ravi Link was constructed having a length of 101 km to add additional water in Ravi River from the Chenab River. In the Sindh Province, construction of Guddu Barrage on the Indus River started in 1957 and was completed in 1963. The aim of this project was to remodel the upper inundation canals in the Sindh area into perennial canals to increase the area under cultivation in Sindh and Balochistan, and it was designed to keep 1.13 million hectares of land irrigated throughout the year. Later on, Kotri Barrage was completed to ensure the supply to the inundation canals in the southern parts of Sindh. Both the Guddu and Kotri Barrage accounted for the conversion of a large deserted area into irrigational lands. A multipurpose barrage, Taunsa Barrage[1], on the Indus River was completed in 1958 to provide controlled water supplies for irrigation [21]. The Warsak Dam[2] was constructed in 1960, and along with providing water storage, it also produced 40 MW of electricity. Later on, the production capacity of the dam was increased by installing additional generators [23].

3. The Indus Water Treaty (IWT) and hydrological developments

In 1960, IWT was signed between Pakistan and India as an effort for resolving the disputes due to the partition of the Indus basin rivers. The headwater sources of Indus River and its tributaries are present in India. India got the control over water resources of Pakistan flowing downstream. During the first 10 years of independence, Pakistan experienced severe blockage or reduction of river waters, which badly affected the crop yield in Pakistan. The World Bank and British Government helped the twin states to reach an agreement called Indus Water Treaty (IWT). In this treaty, the water rights of eastern rivers, viz., Sutlej, Ravi, and Beas Rivers, were allocated to India, whereas the control of western rivers, viz., Indus, Jhelum, and Chenab Rivers, was given to Pakistan. After a period of 10 years, both the countries were authorized to utilize their share of water in their own way. After the treaty was signed, India started to construct projects on eastern rivers in order to divert the water flowing in eastern rivers, and water shortage started in the areas of Pakistan irrigated by the eastern rivers. To fulfill the shortage of water in the eastern rivers, the World Bank financed 8 billion US dollars project "Indus Basin Development Fund" for the construction of dams, barrages, and canals in Pakistan. Furthermore, link canals were constructed to inter-connect the western and the eastern rivers in Punjab for a sustainable supply of water to the cropland. The project was completed in two phases due to the inadequacy of funds. In the first phase, the key construction of Mangla Dam was completed on the Jhelum River in 1967.

Some of the barrages and canals were also modified and improved. After the completion of the first phase, another amount of 1.2 billion US dollars was approved by the World Bank, and the second phase of development started in 1968. The construction of Tarbela Dam on the Indus River was completed in 1976. This dam has sufficient storage capacity of water to supply during low flow season [34]. Both Mangla and Tarbela dams accounted for the major proportion of hydroelectricity generated in Pakistan. Several barrages were remodeled to divert the water from one river to another river such as Rasul, Sidhnai, Chashma, etc. Punjab and Sindh Governments are focusing on lining the canals and water courses to conserve the water. A number of projects have been completed for electricity generation like Ghazi-Barotha and Neelum-Jhelum hydroelectric projects [35]. Owing to the high demand of irrigation water and climate change, rivers in Pakistan are facing the reduction in water flow, and this shortage of water may be intensified in the near future. Considering these issues, the Government of Pakistan is planning to construct more dams to increase the water storage capacity on the Indus River. Details of the existing dams, barrages, link canals, and irrigation canals are given in **Table 2**.

India also carried out some projects in the Indus basin, like Harike Barrage with three canals, viz., Ferozepur Feeder Canal, Makhu Canal, and Rajasthan Feeder Canal, constructed on the Sutlej River in 1952. The Rajasthan Feeder Canal was later on

[1] Barrage is a weir-controlled system installed to divert and control the water flow into the canals with negligible water storage capacity.

[2] Dam is a high concrete walled structure built to store a large volume of water for agriculture and to use the potential energy of water for electric power generation.

River	*Dam(s) (Million m3)*	*Barrages/Head Works/Dam (No. of off taking Canals)*	*Link Canal(s) (Thousand Cusecs)*	*Irrigation Canal(s) (Thousand Cusecs)*
Indus	Nimoo Bazgo (120)	Jinnah Barrage (1)		Thal (6.0)
	Tarbela (11,900)	Chashma Barrage (2)	Chashma-Jhelum (21.7)	Chashma Right Bank (2.5)
		Taunsa Barrage (4)	Taunsa-Panjnad (12.0)	D.G Khan (8.8), Muzaffargarh (7.3), Kacchi (6.0)
		Guddu Barrage (4)		Thar (10.0), Ghotki Feeder (8.5), Pat & Desert Feeder (6.3,12.5), Beghari Feeder (15.3)
		Sukkur Barrage (7)		Khairpur West (1.9), Rohri (11.2), Khairpur East (2.7), Nara (13.4), Dadu (3.2), Rice Feeder (10.1), North West (5.1)
		Kotri Barrage (3)		Kaliari (9.0), Pinyari (14.4), Fuleli (13.8), Lined Channel (4.1)
Jhelum	Mangla (9,000)	Mangla Dam (1)	Upper Jhelum (12.0)	Lower Jhelum (19.0)
		Rasul Barrage (2)	Rasul-Qadirabad (19)	Upper Chenab (7.8), Bombanwali Ravi Bedian Dipalpur (3.1)
Chenab	Salal (285)	Marala Barrage (2)	Marala-Ravi (22.0)	Lower Chenab (16.0)
	Baglihar (395)	Khanki Barrage (1)		
		Qadirabad Barrage (1)	Qadirabad-Balloki (14.5)	Haveli (5.0), Rangpur (2.5)
		Trimmu Barrage (3)	Trimmu-Sidhnai (11.0)	Panjnad (9.0), Abbasia (1.1)
		Panjnad Barrage (2)		
Ravi	Ranjit Sagar Dam (3280)	Madhopur Headworks (2)	Madhopur-Beas (10.0)	Upper Bari Doab (8.2), Kashmir (0.8)
		Balloki Barrage (2)	Balloki-Sulemanki (15.4)	Lower Bari Doab (7.4)
		Sidhnai Barrage (2)	Sidhnai- Mailsi (10.1)	Sidhnai (4.0)
Beas	Pandoh (41)	Pandoh Dam (1)	Beas-Sutlej (16.7)	
	Pong (8,570)	Pong Dam (3)		Pong Left Main (12.5), Mukerian Hydel Channel (11.5), Kandi (0.5)
Sutlej	Bhakra (9,620)		Sutlej Yumna (6.5-7.4) *	
	Nangal (20)	Nangal Dam (2)		Bhakhra Main Line (12.4), Nangal Hydel Channel (14.5)
		Rupar Headworks (2)		Sirhind (12.6), Bist Doab (1.4)
		Harike Headworks (2)		Sirhind Feeder (5.2), Rajasthan Feeder/Indira Gandhi (18.5)
		Ferozepur Headworks (2)		Ferozpur Feeder (11.2), Abohar Branch Lower (1.7)
		Sulemanki Barrage (3)		Upper Pakpatan (6.6), Eastern Siddiqia (4.9), Fordwah (3.4)
		Islam Barrage (2)		Qaim (0.6), Bahawal (5.4)

*Incomplete Project

Table 2.
Hydrological developments (dams, barrages, headworks, link canals, and irrigation canals) on Indus River system (IRS) [4, 27–33].

upgraded in 1961 with the construction of Bhakra Main Line Canal and a large network of distributary canals under Indira Gandhi Irrigation Canal system to irrigate the Rajasthan desert [36]. In 1954, Ravi-Beas Link Canal was designed to transfer the water from Ravi River to Beas River. At the time of partition, the Bhakra Dam was in progress and completed in the early 1970s under the Bhakra Nangal Project. Bhakra Main Line Canal was constructed from Bhakra Dam for irrigation purposes. Downstream of Bhakra, at a distance of 13 km, another dam called Nangal Dam, was designed to control to feed the Nangal Hydel Channel. The whole project generates about 1325 MW of electricity and provides irrigation water for about 4.04 million hectares of land in Punjab, Himachal Pradesh, Haryana, and Rajasthan [37]. The Pong Dam was built in 1975 on Beas River to store water and use it through Shah Nahr. In 1977, Pandoh Dam was constructed along with the Beas-Sutlej Link Canal from the Beas River to Sutlej River. Sutlej-Yumuna Link Canal started from Nangal Dam in 1982, but it is still incomplete due to strong resistance from the Indian Punjab Government [33].

In 2014, Nimoo-Bazgo Dam was completed on Indus River in Ladakh region of Indian-held Jammu and Kashmir. Similarly, Kishanganga Dam is under construction by India, on the upstream of Neelum River near *Bandipore*, to divert the water flow into an underground powerhouse to generate electricity [38]. Besides these projects, a large number of projects are still in planning stages on both sides of the borderline. Majority of these projects are of hydroelectric nature, and water will mainly be used for electricity generation, for example, Diamer-Bhasha Dam and Mohmand Dam, from Pakistan's side and Pakal Dul Dam from the Indian side. But with the construction of these massive concrete structures, a huge amount of water will be stored and provided during low water season in order to fulfill the agricultural water requirements.

4. Socioeconomic impacts on local population

Early irrigation system in the Indus basin was developed for irrigational purposes. Due to the development of canals in the Kashmir region, the agricultural production was increased many folds [39]. The Indus basin irrigation system is one of the largest, complex irrigation networks and the largest weir-controlled system in the world with large reservoirs, barrages, syphons, and many types of canals and their watercourses [40]. Construction of new canals created favorable conditions for the colonies to be developed in the basin. Best examples can be seen as Lyallpur (Faisalabad) developed in the vicinity of the Lower Chenab Canal. Similarly, Sargodha and Montgomery (Sahiwal) were developed along the lower Jhelum Canal and the Lower Bari Doab Canal, respectively [25]. High crop productivity was found in irrigated areas than in the rainfed areas. High crop production leads to high per hectare employment. Additionally, irrigation system also made it possible for an extra crop to be raised, hence increasing per year yield, increasing household income, and decreasing poverty in irrigated areas [41].

Development of irrigation system benefited both at the local and governmental levels. A famine-prone area was converted into an area with high grain productivity that helped to prevent the famine conditions. Living style of the inhabitants was improved, and starvation was minimized. At the same time when the common population was benefitted, the British Government harvested a large volume of cotton crops from the irrigated areas, which provided them with a golden opportunity to cash the results of irrigation system development. Furthermore, the railway network increased access to agricultural products. Railway network facilitated the transport of food grains and cotton to local and international markets. Food grains produced from the area also helped to meet the food demands of other regions, hence making the region of South Asia's "breadbasket." Without the irrigation system, it seemed to be impossible because a large area was just barren and unproductive due to insufficient irrigation water. Similarly, livestock transformed from nomadic form to an organized industry to rear animals for milk and meat production as more fodder crops were available.

Construction of dams provided a platform for the production of electricity from river water. In order to fulfill the commercial and domestic electric requirements in the region, turbines were installed to produce the hydroelectric power. Industrial production increased due to the availability of a cheaper electric supply subsequently increasing the revenue from the industrial sector. Agricultural-based industrial development produced a large number of jobs to boost the local economy. Poverty was reduced, and livelihood increased favoring a better lifestyle in the families living near industrial areas.

5. Water sharing conflicts

At the time of partition of the subcontinent, the division took place in such a way that Kashmir became a territory of common interest being an origin for the most of rivers flowing downward to Pakistan. Additionally, India cut off the water flowing into all the canals of Pakistan originating from the headworks situated in India, which created a situation of restlessness among the people of both countries. In order to minimize the tension created between both countries over the river water issue, IWT was signed with the help of the World Bank [6, 42]. This agreement created an atmosphere of competition to build dams and barrages to not let their share of water be wasted. Especially, India constructed many dams and barrages in the upstream area of Jammu and Kashmir and on Rivers flowing through Indian territory, i.e. Ravi, Beas, and Sutlej. Due to the construction of these massive structures and diversion of water, a lot of detrimental effects have been produced in the Indus basin [43]. Construction of storage reservoirs on Indus River tributaries by India has lowered the downstream water flow creating a state of political unrest between India and Pakistan. Afghanistan is also planning to build dams on the Kabul River which may generate a new political issue between Pakistan and Afghanistan [40].

A hydrological disagreement between India and Pakistan has always been there since the very first instance being the construction of Salal hydroelectric project by India on Chenab River. Initially, Pakistan disagreed to the project, but after negotiations, the project was accepted in the 1970s. Afterward, many projects were subjected to disagreement including the Wullar Barrage and 430 MW Baglihar hydroelectric project. In the 1980s, India stopped the Wullar project due to extreme opposition from the Pakistan side [44]. In 1999, Pakistan complained about the second project, but negotiation continued on many forums. In 2007, after the acceptance of some modification suggested by Pakistan, the project was allowed to be completed [45]. Issues were also raised by Pakistan on the construction of Uri Dam on Jhelum River (1997) and Ratle Project on Chenab River (2013). The Ratle project consists of construction of a 170 m concrete wall to produce 850 MW electricity [46]. Uri dam was constructed during 1989–1997 with an installed capacity of 480 MW, but another project was run by India in 2005 to increase the production capacity of the Uri Dam. This Uri-II project increased the capacity by 240 MW through the construction of 3.61 km long tunnel with a diameter of 8.4 m [47]. In spite of the opposition from Pakistan's side, the project was completed in 2014. Recently, India started the construction of 330 MW Kishanganga hydroelectric project in 2007 which is considered as the most controversial project of the history till now. By the construction of this project, water flow in the Neelum River will become low and may cause failure of 969 MW Neelum-Jhelum project that has to be constructed downstream by Pakistan [44]. According to the opinion of John Briscoe, a water expert from South Africa, if all the projects designed by India are successfully completed on the Indus River, it will destruct the agricultural crops of the dry season in Pakistan as India will be able to hold up a month's worth river supply in these reservoirs. Unavailability of water will make the plantation difficult and will have negative effects on the agricultural sector of Pakistan [48]. Not only Pakistan is being affected, but in the Kashmir region, India is also producing electricity for its own provinces at the expense of only 12% given to Kashmir itself as a royalty. To fulfill the needs of electricity generation in the peak hours, Kashmir has to purchase the electricity generated from its own waters at an inflated rate despite having a production potential for 20,000 MW of electricity. At the same time, when IWT tried to resolve the water tension between Pakistan and India, it also opened the doors of discrimination for the disputed region of Kashmir because the agreement was signed without the consultation of Kashmir's prime minister [43].

After the partition, India started the construction of barrages and dams on eastern rivers, to divert the water flow to its own states rather than to Pakistan. Through the construction of Sutlej-Yumuna Link (SYL) Canal, a large portion of irrigational water (3.45 MAF) was planned to be moved from the Indian Punjab province to Rajasthan and other provinces. It was designed to irrigate about 446,000 ha land in Haryana and 128,000 ha in Punjab [49]. Inhabitants of Punjab claim that this diversion was not justifiable as it will turn irrigational areas of Punjab into desserts due to water scarcity. Nowadays, like most areas of the basin in Pakistan, the areas of Indian Punjab have become water scarce, and the water table has been lowered to a drastic level due to less availability of surface water. Farmers are forced to extract the groundwater which adds to the irrigational costs and negatively affects the income of the agricultural sector [50]. Water flow in the Sutlej River downstream of Harike Headworks is almost zero, and it is only flowing in case of high floods or due to any failure in Bhakra or Nangal Dam. It makes the downstream areas of Punjab dry during many parts of the year and flood-prone due to storage of large water in the upstream reservoirs. Construction of SYL is still incomplete due to interprovincial politics and concerns over the division of water [51].

Similarly, interprovincial conflicts also exist in Pakistan over the unequal distribution of Indus waters among the dominant province of Punjab and other smaller provinces mainly Sindh [52]. Unlike India, these conflicts have been limited to the political platforms. In the mid-nineteenth century, the construction of large canal structures by the British Government in Punjab gave birth to these conflicts. In 1945, Interprovincial water distribution was ensured by signing a treaty in which the Indus River water along with its tributaries was distributed between Sindh and Punjab. According to this treaty, majority of water from the eastern tributaries of Indus (94%) were allocated to Punjab and remaining to the Sindh. Meanwhile, Sindh was allocated with 75% water of the Indus main channel and Punjab with the remaining 25%. The Indus Water Treaty made the construction of link canals necessary for Pakistan in order to compensate for the upstream loss of water. As a majority of the link canals and storage reservoirs were to be constructed in the Punjab region, Sindh's population perceived it as a conspiracy to compensate Punjab at the expense of Sindh's share of the Indus River [45].

To settle the issues among the provinces, the "Water Accord 1991" was endorsed in which Indus water was distributed on the basis of average flow. Even after the accord, some insecurities existed in smaller provinces because the large quantity of water was allocated to only Punjab and Sindh [52]. Kalabagh Dam has been a focus of attention for many years due to political insecurities on the construction of the dam. People of Sindh claim that after the construction of the dam, water flow will be reduced causing droughts and saltwater intrusion in downstream areas. They also claim that Sindh is far more dependent on river waters because more than 80% of groundwater is saline in the region and construction of the dam will compromise their water requirements. This project was developed in the light of past destructive events of droughts in Sindh and subjected to opposition from politicians and bureaucrats of the Sindh province [45].

6. Ecological and environmental adversities

Since the prehistoric era, changes in irrigation system have a strong link with climate change [39]. Due to the increase in agricultural area and productivity, the human population has been increased in the Indus basin, followed by urbanization and industrialization, boosting the water demands for household, food production, and energy and industrial sectors. Population increased due to the availability of resources and settlement areas. A large area was converted into the urbanized area, and some of the native species vanished from the Indus Basin. In order to protect the urban area, a natural path of the Indus River should be restored along with its natural floodplains and wetland area. The presence of highly populated areas in the floodplains of rivers makes a large number of people prone to the flooding. In such circumstances, more economic loss has been observed by the flood as it happened in 2010 in Pakistan. In Kashmir region, the majority of the dams are constructed in a hazard-prone area and are called as "water bombs" by a glaciologist. About 15 dams were built by the Indian Government in the Himalayas, which were not recommended due to fragile mountainous land. The temperature in this mountainous region is increasing by human-induced global warming, and glaciers are retreating, creating a temporarily high flow in the Indus River, which may cause damage to the structures built [43]. The sediment load is constantly increasing in the river resulting in avulsion. With the passage of time, the storage capacity of the reservoirs started to decline due to the siltation increasing the area under inundation [53]. Reduction in the storage capacity increases the chances of flooding in the surrounding areas in high flow seasons. A detailed information about the impacts on the Indus River and mitigation measures is provided in **Table 3**.

Problem	Affected area(s)	Mitigation measure(s)	Recommendation(s)
Waterlogging and Salinity	Lower and middle Indus basin	Salinity Control and Reclamation Project	Changing irrigation techniques, using gypsum
Habitat fragmentation	Indus Blind Dolphin	Indus River Dolphin Conservation Project	Construction and proper maintenance of fish ladders, preventing low water level
Increased flooding	Lower floodplain areas	Construction of dikes, Flood prediction and migration	Increasing storage capacity, emergency response plan
Riverine pollution	Fish species, water quality	Solid waste management, Industrial and municipal wastewater treatment	Control over fertilizer and Pesticide usage, Organic farming
Groundwater extraction	Punjab and Sindh	Water re-use and conservation, Drought resistant crops	Rain water harvesting, Groundwater recharging, Increasing storage capacity, Awareness campaigns
Declining delta and estuaries	Mangroves	Mangroves restoration projects	Construction of levees along the coastline, mangroves plantation, increasing water flow, reducing mangroves burning as fuel

Table 3.
Environmental and ecological problems in Indus Basin, their mitigation and recommendations [53–62].

Since 1851, the salinity problem has been observed in different areas of the Indus basin. Firstly, the salinity issue was identified in the Jammu and Kashmir regions. Similarly, in Punjab, salinity problems were also reported. To check the overall groundwater level in the basin, a series of observatory wells were installed across the basin. In open water table wells, Punjab showed the remarkable increase in the water table level induced by man-made irrigation system in the area. The main reason for the high level of the water table, salinity, and water logging observed in the basin, although it was not that much prominent, was considered to be the diversion of the river channel and the construction of unlined irrigation canals during 1850–1950. The situation in Punjab is still considered better than Sindh because in this area water logging was more common than salinity due to salt-free upper layers and better drainage and topography. Salinity problems in Punjab were only confined to the areas with poor drainage and topography. In the central Indus basin, i.e., lower Punjab and upper Sindh, the water table was found to be highest in the 1940–1950s followed by the highest level of waterlogging in the 1960–1980s. Salinity conditions in this area are in transition to the upper and lower areas of the basin. The productivity of 20–30% of irrigated land was affected due to salinity and waterlogging during 1970–1980s [63].

A shift in climatic patterns and over the use of surface water reduced the availability of water, forcing the people to rely on groundwater aquifers, resulting in rapid depletion of subsurface water resources. The shift from surface irrigation to groundwater is the inability of the Indus basin irrigation system to accommodate the changing water requirements of crops over different seasons due to inefficient water management. Increase in food requirements due to population growth and economic competition emerged due to over demand of crop yield, and the farmer community moved from a conventional irrigated agriculture to more water-intensive agriculture that was beyond the capacity of the existing irrigation system. More than 80% of groundwater is extracted by small tube wells because of cheaper installation and easy operation. In the irrigated areas, groundwater is of poor quality, and its frequent use in agricultural has resulted in salinity of a large portion of the agricultural area in the Indus basin [53].

The Indus River delta is facing the high risk of salinity in upper parts due to vertical and lateral movement of saline groundwater into the fresh shallow aquifers [64]. Secondary salinization has been observed in 4.5 million hectares of land, and half of the lands are in the Indus basin irrigation area. About one million hectares of agricultural land is facing waterlogging problems due to seepage of canal water and poor irrigation techniques [45]. Food grain crops have been replaced with high-price crops, which require more water than the food grains, resulting in increased water demand. The canal water system, fed by Indus River, adds almost 16.6 million tons of salts into the irrigated and deserted strata of Indus basin including groundwater aquifers. Rainstorms cause a lot of agricultural and economic damages due to the restricted surface and subsurface water drainage, because of the plain topography of Indus basin [40].

Development of irrigation system structures such as dams and barrages has led to the fragmentation of Indus River into 17 sections which resulted in extirpation of dolphins from 10 sections. They were found to be present in only 6 out of 17 sections of the river due to the usage of river water for irrigation and low water discharges in the dry season. Spatial and temporal distribution patterns were affected by habitat fragmentation, and, combined with habitat degradation, it contributed to the decline of the dolphin population [55]. The whole length of the Indus River used to be the habitat of dolphins, but this area has now been restricted to only 20% of total length due to habitat fragmentation. Chemical pollution and accidental death by fishing gears are some of the potential factors for the decline in the dolphin population [65]. Sometimes, dolphins get trapped in the canals, fail to return back into the river, and eventually die. In the downstream area of Sukkur Barrage, the river channel is highly constricted, and dolphins are subjected to intensive fishing activities mainly in the winter season [66].

Natural processes have the potential to contaminate water resources, but with the development of irrigation system and residential colonies, the agricultural system became widespread, and a large number of industries came into existence. Anthropogenic activities in agriculture and industries cause discharges of fertilizers, pesticides, chemicals, heavy metals, and pathogens that can degrade the water quality and can cause negative effects on human health. Fertilizers and pesticides used in the agricultural areas are washed off and drained into surface and groundwater aquifers. The water requirements are high in the urban areas, and they contribute to water pollution by the production of municipal sewerage and leaching from solid waste generated. Water affected both qualitatively and quantitatively making it unsuitable for the human and animal consumption. As the Indus River flows downward, the effects of deterioration are intensified as a large volume of untreated effluents from agricultural, industrial, domestic, and commercial areas enter into the river on its way. Throughout the river channel, concentrations of nitrogen, phosphorus organic matter, pesticides, and mercury are present at an alarming level. Most of the effluent dumped into the river comes from the agricultural sector alone, while industries, households, and urbanized areas being the other major sources [67].

Indus delta is rich in biodiversity, and mangroves were a very important ecological resource in the lower Indus basin. Reduced water flow, pollution of rivers, and usage of mangroves as fodder and fuel have reduced the mangrove species. The population of mangroves has declined to a threatened level by pollution and anthropogenic activities [68]. With the decline in mangrove population, the spawning and rearing sites of many fish and macroinvertebrate species have been destroyed causing a decline in the population of those species. The flow of water in Sutlej River has almost gone to zero due to upstream storage by India. The river channel has been converted into a sandy desert with no water at all losing all the scenic values associated with its original aquatic ecosystem. As the river dried up, agriculture declined, and biodiversity was rapidly declined which disturbed the regional ecology. People moved from the dry areas, and all the developed infrastructure was destroyed. Sutlej Valley developed because the river began to die as the people migrated out of the area due to perished livelihood. Much of the culture of this valley has already been lost, and if not managed for a few more years, the civilization will be completely vanished [69].

7. Recommendations and conclusion

No doubt the Indus irrigation system is one of the largest irrigation systems in the world. It provides food, jobs, and recreational resources for Pakistani and Indian people, but in the race of extensive water use, the environmental resources of Indus basin have been compromised. Natural flow and discharge of rivers have been altered by the construction of massive structures. At the beginning of irrigation system development, the groundwater level raised, and it changed the vegetation pattern in the region along with the intensive agricultural activities favored by a large supply of irrigation water. Large-scale agricultural activities attracted the human population to be settled in this region, thus making it one of the thickly populated regions in the world. This situation created an environmental disaster in the region. Indiscriminate human activities contaminated surface and groundwater resources through industrial and agricultural chemicals. Several types of diseases and human abnormalities have been reported from this region due to excessive use of the chemicals in the urban and rural areas. The Indus River, once regarded as the Mighty Indus, has already turned into an ordinary polluted river. The situation is now or never, and we have to take some concrete measures to restore the natural flow of the river and develop efficient technologies to get maximum crop yield with minimum use of water. The sustainable approach could be beneficial for the future of rivers and human population.

After 150 years of the development of irrigation systems, most of the environmentalists come up with a conclusion that sustainable environmental management of rivers must be ensured. The authorities should maintain minimum ecological flow downstream to sustain the ecological conditions of river. Fish and other organism communities have been constantly declining due to decreasing water level and fragmentation of their habitat. Further pressure has been exerted on the fish population by intensive fishing and using illegal fishing nets. Policies must be implemented by keeping in view the maintaining of the minimum flow of water in river to support the ecological health of Indus River system. To avoid habitat fragmentation, upstream and downstream habitats should be connected through functional fish ladders. Fishing activities must be monitored regularly, and fishermen community should be encouraged to use legal gears for fishing activities.

Human activities in the catchment area and indiscriminate use of water from Indus River system are at full swing on both sides of Pakistan and India. Due to intensive agriculture, the river water is becoming insufficient to meet the irrigational demands despite a well-established irrigation system. Groundwater extraction is carried out for compensating the shortage of surface water resulting in the lowering of groundwater table. Increasing urbanization and development of industrial sector enhanced the utilization of subsurface water for the human consumption and industrial manufacturing. Groundwater aquifers have been exhausted down to the depth of several hundred feet in majority of the areas of Indus basin. Both countries must sign a new agreement for the stringent control of the noxious human activities. They must work on drought-resistant crop varieties to ensure the minimum use of surface and groundwater for irrigation purposes.

River pollution is one of the major threats to the ecological resources. It causes substantial damages to the fauna and flora of aquatic ecosystem. Industries use many types of chemicals in their manufacturing processes, and majority of water used to wash off the chemicals is drained into rivers without proper treatment. Commercial and residential areas also generate a large amount of municipal sewerage water that finds its way to the river without any treatment. Industrial and municipal waste water treatment should be ensured with strict rules and regulations. Continuous monitoring of treatment plants should be ensured to keep them in a proper working condition.

Pakistan and India are in a developmental race to take control over the Indus water resources in order to improve their economy and to minimize the shortfall of electricity in the region. Construction of massive hydroelectric projects and diversion channels has adversely affected the natural river course. Currently, it is impossible to predict the situation of Indus River in the upcoming years when the water resources are being exhausted rapidly. Sustainable use of water resources can be helpful in minimizing the ecological damages at the expense of economic development. Both countries should modify their developmental policies by limiting the human interventions in the natural course of rivers. In this way river system's natural condition could be sustained.

Considering the pace of hydrological development on the Indus River System, Integrated Water Resource Management (IWRM) must be implemented especially in Pakistan as it is the only water source of the country [70]. It is important to efficiently utilize all the available surface and groundwater. Optimum amount of water should be allocated to each sector and area to reduce the pressure on groundwater resources. Farmers can be educated to use the modern agriculture practices and drought-resistant crops in order to save water. Infrastructure for the storage of water can be enhanced by the government to reduce the impacts of low water season on the economy of agricultural community. Restriction on the excessive usage of pesticide and fertilizers on crops should be enforced to avoid water pollution. Moreover, it has been observed that Indus Water Treaty was the need of time, but it needs a lot of improvements to meet the environmental issues of changing climate and rising tensions between India and Pakistan. Both countries must negotiate to bring some new provisions in the treaty to maintain a minimum amount of water in the river system necessary for the ecological requirements of the basin. At this stage, if the Indus River system is not managed sustainably, the system may undergo irreversible damages that can never be undone.

Acknowledgements

The authors acknowledge the services of Mr. Irfan Ali in map preparation included in the chapter.

Conflict of interest

The authors hereby declare that there is no conflict of interest.

Author details

Muhammad Irfan[1], Abdul Qadir[1*], Habib Ali[2], Nadia Jamil[1] and Sajid Rashid Ahmad[1]

1 College of Earth and Environmental Sciences, University of the Punjab, Lahore, Pakistan

2 Pakistan Science Foundation, Islamabad, Pakistan

*Address all correspondence to: aqadir.cees@pu.edu.pk

References

[1] Indus Basin [Internet]. 2013. Available from: https://www.waterinfo.net.pk/?q=node/550 [Accessed: 2018-12-19]

[2] Indus River [Internet]. 2016. Available from: https://www.britannica.com/place/Indus-River [Accessed: 2018-12-09]

[3] Gupta A. Large Rivers: Geomorphology and Management. Wiley; 2008. 712 p

[4] Stewart J, Podger G, Ahmad M, Shah M, Bodla H, Khero Z, et al. Indus river system model (IRSM)—A planning tool to explore water management options in Pakistan: Model conceptualisation, configuration and calibration. In: Technical Report. South Asia Sustainable Development Investment Portfolio (SDIP) Project. Australia. 155p EP186945: CSIRO; 2018

[5] Case Study 3: The Indus River Basin [Internet]. 2007. Available from: http://www.caee.utexas.edu/prof/mckinney/ce397/Readings/case-study3_Indus.pdf [Accessed: 2019-04-12]

[6] Biswas AK. Indus water treaty: The negotiating process. Water International. 1992;**17**(4):201-209

[7] Alam UZ. Questioning the water wars rationale: A case study of the Indus waters treaty. Geographical Journal. 2002;**168**(4):341-353

[8] The Indus Basin—An Overview [Internet]. 2017. Available from: https://slideplayer.com/slide/9728703/ [Accessed: 2019-04-12]

[9] Transboundary Water Cooperation in the Indus Basin: Challenges & Opportunities [Internet]. 2017. Available from: https://www.gwp.org/globalassets/global/gwp-sas_images/gwp-sas-in-action/transboundary/dr-shaheen-akhtar.pdf

[10] Regional Workshop on Transboundary Water Cooperation in the context of the SDGs in South Asia and beyond [Internet]. 2017. Available from: https://www.gwp.org/contentassets/9047146502d64c8889ec57b838ba7e9e/transboundary-cooperation-report-final-draft.pdf [Accessed: 2019-04-15]

[11] International Rivers of Western Himalayas—Hydrological Prospects and Areas of Concern [Internet]. 2019. Available from: https://www.slideserve.com/sook/international-rivers-of-western-himalayas-hydrological-prospects-and-areas-of-concern-powerpoint-ppt-presentation [Accessed: 2019-04-22]

[12] Jain SK, Agarwal PK, Singh VP. Hydrology and Water Resources of India. Springer Science & Business Media; 2007. 1260 p

[13] Major Rivers Flowing in Punjab [Internet]. 2015. Available from: http://www.india-wris.nrsc.gov.in/wrpinfo/index.php?title=Major_Rivers_Flowing_in_Punjab [Accessed: 2019-05-04]

[14] Vorosmarty CJ, Fekete BM, Tucker BA. Global River Discharge, 1807-1991, V. 1.1 (RivDIS). Oak Ridge, Tennessee, USA: ORNL DAAC; 1998. DOI: 10.3334/ORNLDAAC/199

[15] McIntosh J. The Ancient Indus Valley: New Perspectives. ABC-CLIO; 2008. 441 p

[16] Mountjoy S, McNeese T. The Indus River. Philadelphia: Chelsea House Publishers; 2004. 117 p

[17] Alam U, Sahota P, Jeffrey P. Irrigation in the Indus basin: A history of unsustainability? Water Science & Technology: Water Supply. 2007;**7**(1):211-218. DOI: 10.2166/ws.2007.024

[18] Wheeler M. The Indus Civilization. 3rd ed. Cambridge University Press; 1968. 143 p

[19] Harappan Hydro-Engineering and Water Management [Internet]. 2011. Available from: www.tifr.res.in/~archaeo/FOP/.../Bisht%20Hydro-Engineering.pdf [Accessed: 2019-02-16]

[20] Kuchy AA. Irrigation works and agricultural development of Kashmir during Avantivarman's reign (855-883 AD). International Journal of Advanced Research in Management and Social Sciences. 2017;**6**(2):27-30

[21] ICID. Irrigation & Drainage in the World—A Global Review [Internet]. 2014. Available from: https://www.waterinfo.net.pk/?q=node/718 [Accessed: 2018-12-13]

[22] Wahi T. Shah Nahr: Its history, technology and socio-political implications. Proceedings of the Indian History Congress. 2013;**74**:285-296

[23] Gilmartin D. Blood and Water: The Indus River Basin in Modern History. Oakland, CA: University of California Press; 2015. 356 p

[24] Naqvi SA. Indus Waters and Social Change: The Evolution and Transition of Agrarian Society in Pakistan. Pakistan: Oxford University Press; 2012. 60 p

[25] Bagchi KS. Irrigation in India: History & Potentials of Social Management. Delhi: Upalabdhi, Trust for Developmental Initiatives; 1995

[26] Sukkur Barrage [Internet]. Available from: http://sindh.gov.pk/dpt/Irrigation/sukkur%20barrage.htm [Accessed: 2018-12-15]

[27] Geopolitics-Water Availability in Pakistan [Internet]. 2016. Available from: http://www.indiandefencereview.com/news/water-availability-in-pakistan/2/ [Accessed: 2019-04-23]

[28] Indus Basin [Internet]. 2011. Available from: http://www.fao.org/nr/water/aquastat/basins/indus/index.stm [Accessed: 2019-04-23]

[29] Canal administration system [Internet]. Available from: http://irrigation.punjab.gov.in/OldVersion/canal_administration_cansystem.html [Accessed: 2019-04-23]

[30] Punjab ENVIS Centre Newsletter. 2015;**12**(3) [Internet]. Available from: http://www.rcenetwork.org/portal/sites/default/files/Envis%20Newsletter%Vol%2012%20No.3.pdf [Accessed: 2019-04-29]

[31] Mukerian hydel power project to start from December [Internet]. 2015. Available from: https://www.thehindu.com/news/national/other-states/mukerian-hydel-power-project-to-start-from-december/article7059526.ece [Accessed: 2019-04-23]

[32] Farmers wait as second phase of Kandi canal yet to get operational [Internet]. 2014. Available from: https://www.hindustantimes.com/punjab/farmers-wait-as-second-phase-of-kandi-canal-yet-to-get-operational/story-k3iptGxwhav0v6mZAKgEcI.html [Accessed: 2019-04-25]

[33] Sutlej-Yamuna Link: Dispute Continues for 61 Years after 1st Agreement in 1955 [Internet]. 2016. Available from: https://economictimes.indiatimes.com/news/politics-and-nation/syl-dispute-continues-for-61-years-after-1st-agreement-in-1955/articleshow/55355851.cms [Accessed: 2019-04-25]

[34] Pakistan Irrigation [Internet]. 1994. Available from: http://www.country-data.com/cgi-bin/query/r-9832.html [Accessed: 2018-12-13]

[35] 4th Neelum-Jhelum Hydropower Project unit begins operation [Internet]. 2018. Available from: https://www.dawn.com/news/1426837 [Accessed: 2019-02-16]

[36] Varis O, Tortajada C, Biswas AK. Management of Transboundary Rivers and Lakes. 1st ed. Berlin Heidelberg: Springer; 2008. 304 p. DOI: 10.1007/978-3-540-74928-8

[37] Bhakra Nangal Project, India [Internet]. 2015. Available from: https://ejatlas.org/conflict/bhakra-nangal-project-india [Accessed: 2018-12-15]

[38] Kishanganga dam issue: World Bank asks Pakistan to accept India's demand of 'neutral expert' [Internet]. 2018. Available from: https://timesofindia.indiatimes.com/india/kishanganga-dam-world-bank-asks-pakistan-to-accept-indias-demand-of-neutral-expert/articleshow/64466122.cms [Accessed: 2019-02-15]

[39] Wescoat JL Jr. Managing the Indus River basin in light of climate change: Four conceptual approaches. Global Environmental Change. 1991;**1**(5):381-395

[40] Qureshi AS. Water management in the Indus basin in Pakistan: Challenges and opportunities. Mountain Research Development. 2011;**31**(3):252-260. DOI: 10.1659/MRD-JOURNAL-D-11-00019.1

[41] Hussain I, Hanjra MA. Irrigation and poverty alleviation: Review of the empirical evidence. Irrigation and Drainage. 2004;**53**(1):1-15. DOI: 10.1002/ird.114

[42] Begum T. Indus water treaty & emerging water issues. Abasyn University Journal of Social Sciences. 2011;**4**(2)

[43] Kashmir: A Water War in the Making? [Internet]. 2016. Available from: https://thediplomat.com/2016/06/kashmir-a-water-war-in-the-making/ [Accessed: 2018-12-19]

[44] Baglihar and Kishanganga: Problems of Trust [Internet]. 2005. Available from: http://www.ipcs.org/comm_select.php?articleNo=1762 [Accessed: 2019-03-04]

[45] Hydropolitics in Pakistan's Indus Basin [Internet]. 2010. Available from: https://www.usip.org/publications/2010/11/hydropolitics-pakistans-indus-basin [Accessed: 2019-02-16]

[46] Another blow to Pakistan's water interests...India starts building 850MW Ratle Dam on Chenab [Internet]. 2019. Available from: https://www.thenews.com.pk/archive/print/631633-another-blow-to-pakistan%E2%80%99s-water-interests...india-starts-building-850mw-ratle-dam-on-chenab [Accessed: 2019-03-19]

[47] Uri-II Hydroelectric Project, Jammu and Kashmir [Internet]. 2019. Available from: https://www.power-technology.com/projects/uri-ii-hydroelectric-project-jammu-and-kashmir/ [Accessed: 2019-03-19]

[48] War or Peace on the Indus [Internet]. 2010. Available from: https://johnbriscoe.seas.harvard.edu/files/johnbriscoe/files/108._john_briscoe_war_or_peace_on_the_indus_201004.pdf [Accessed: 2019-03-04]

[49] Sutlej-Yamuna Link Canal [Internet]. 2018. Available from: http://mowr.gov.in/sutlej-yamuna-link-canal [Accessed: 2019-03-04]

[50] Baweja S, Aggarwal R, Brar M. Groundwater depletion in Punjab, India. In: Encyclopedia of Soil Science. Vol. 2017. pp. 1-5. DOI: 10.1081/E-ESS3-120052901

[51] Final Assault—A planned desertification of Punjab. [YouTube].

2016. Available from: https://www.youtube.com/watch?v=-2_Viws3Ang [Accessed: 2018-12-15]

[52] Kanwal L. Sind-Punjab water sharing conflict. Pakistan Journal of Social Sciences. 2014;**34**(2):501-510

[53] Laghari A, Vanham D, Rauch W. The Indus basin in the framework of current and future water resources management. Hydrology and Earth System Sciences. 2012;**16**(4):1063-1083. DOI: 10.5194/hess-16-1063-2012

[54] Alam A. Analysis of Salinity Control and Reclamation Projects. 2015

[55] Braulik GT, Arshad M, Noureen U, Northridge SP. Habitat fragmentation and species extirpation in freshwater ecosystems; causes of range decline of the Indus River dolphin (*Platanista gangetica minor*). PLoS One. 2014;**9**(7). DOI: 10.1371/journal.pone.0101657

[56] Dandekar P. Dolphins of the Ganga: Few Fading, Fewer Frolicking [Internet]. 2011. Available from: https://sandrp.files.wordpress.com/2018/03/dolphins_of_the_ganga-few_fading_fewer_frolicking.pdf [Accessed: 2019-03-25]

[57] Indus River Dolphin Conservation Project [Internet]. 2011. Available from: http://www.wwfpak.org/wwf-projects/indusdolphin.php [Accessed: 2019-03-20]

[58] Ali A. Indus Basin Floods: Mechanisms, Impacts, and Management [Internet]. 2013. Available from: https://www.adb.org/sites/default/files/publication/30431/indus-basin-floods.pdf [Accessed: 2019-03-15]

[59] Polluted water [Internet]. 2017. Available from: https://www.thenews.com.pk/print/229823-Polluted-water [Accessed: 2019-03-29]

[60] Ali Z. Indus delta has shrunk by 92% since 1833. 2018. Available from: https://tribune.com.pk/story/1649523/1-indus-delta-shrunk-92-since-1833/ [Accessed: 2019-03-29]

[61] Restoration and Rehabilitation of Mangroves Ecosystem along the Coasts of Pakistan [Internet]. Available from: https://www.iucn.org/asia/countries/pakistan/restoration-and-rehabilitation-mangroves-ecosystem-along-coasts-pakistan [Accessed: 2019-03-30]

[62] IUCN and Engro Elengy Terminal Limited to restore mangroves on 500 hectares of coastline [Internet]. Pakistan. 2016. Available from: https://www.mangrovesforthefuture.org/news-and-media/news/pakistan/2016/pakistan-iucn-and-engro-elengy-terminal-limited-to-restore-mangroves-on-500-hectares-of-coastline/ [Accessed: 2019-03-30]

[63] Bhutta MN, Smedema LK. One hundred years of waterlogging and salinity control in the Indus valley, Pakistan: A historical review. Irrigation and Drainage. 2007;**56**(S1):S81-S90. DOI: 10.1002/ird.333

[64] Khan S, Rana T, Gabriel HF, Ullah MK. Hydrogeologic assessment of escalating groundwater exploitation in the Indus Basin, Pakistan. Hydrogeology Journal. 2008;**16**(8):1635-1654. DOI: 10.1007/s10040-008-0336-8

[65] Braulik GT, Noureen U, Arshad M, Reeves RR. Review of status, threats, and conservation management options for the endangered Indus River blind dolphin. Biological Conservation. 2015;**192**:30-41. DOI: 10.1016/j.biocon.2015.09.008

[66] Reeves RR, Chaudhry AA. Status of the Indus river dolphin Platanista minor. Oryx. 1998;**32**(1):35-44. DOI: 10.1046/j.1365-3008.1998.00016.x

[67] Iqbal AR. Environmental Issues of Indus River Basin: An Analysis. Vol. 5(1). Islamabad, Pakistan: ISSRA Papers

Institute for Strategic Studies, Research and Analysis, National Defence University; 2013. pp. 89-112

[68] Amjad AS, Kasawani I, Kamaruzaman J. Degradation of Indus delta mangroves in Pakistan. International Journal of Geology. 2007;**3**(1):27-34

[69] Iqbal MJ, Anwar M, Minallah MNU, Chandio NH, Zakharov K, Mohsin M, et al. Spatio-temporal and physiographical study of the abandoned Sutlej river: A case of Jhangi Wala, Bahawalpur, Pakistan. Journal of Basic and Applied Sciences. 2018;**14**:124-130. DOI: 10.6000/1927-5129.2018.14.18

[70] Muhammad Saleem A. Responding water challenges in Indus River basin of Pakistan. 2016

Chapter 6

Spate Irrigation: Impact of Climate Change with Specific Reference to Pakistan

Qudrat Ullah Khan and Obaid Ullah Sayal

Abstract

Spate irrigation is a unique system of agriculture practiced in the piedmont plains by harvesting of floods received after rainfall in the mountains. This system is practiced in different parts of the world; in Pakistan, it is extensive in the western belt. The system is based on water distribution from head to tail. There are laws for distribution of water, but due to the magnitude of flood, it sometimes retains in the upstream and sometimes finds its way to the river. Agriculture practiced in this system depends on floods, which brings sedimentation, useful in replenishing soil fertility. Soil has the ability to hold moisture for long. The changing climatic pattern has greatly influenced the system both under droughts and floods. Livelihood of the spate farmers depends on agricultural crops and livestock. In either case of the extreme climate, they have to cope with limited options. Changing climatic pattern is responsible for extending the climatic seasons and enhancing the irrationality of floods. Construction of huge dam on the torrential watershed is a great project executed by the government for large floods, overcomes energy crisis, and has potential to irrigate land through canal. This chapter is a brief comprehension of spate irrigation under changing climate with special focus on Pakistan.

Keywords: spate irrigation, climate change, Pakistan, water rights

1. Introduction

Spate irrigation is an old system of agriculture practised in the foothill plains. The system has exclusivity as the area faces the two extremes: the drought in the dry season and huge floods in the wet seasons. There are a number of countries in the world where the spate irrigation is practised; the more obvious are Eretria, Ethiopia, Iran, Pakistan and Yemen [1]. The more extensive spate irrigation system is practised in Pakistan. In Pakistan the system extends near the Sulaiman ranges in the western part of the country.

Floods considered as catastrophe are a colossal opportunity for the spate irrigation farmers. The farmers of the area wait till the monsoon and spring floods. The livelihood of the farming community depends on these floods as a source of irrigation and also drinking water. The floods once received bring fortune to the farmers. Spate irrigation in different parts of the world is defined by different people, as discharged flood from mountainous watershed that flows through different channels in agriculture fields [2], whereas others termed spate irrigation as flood water that flows in the torrent beds and stream from the hilly watershed and spread in cultivable fields used for growing the crops [3]. Also some of the researchers have defined spate irrigation as "Diversion and distribution of torrential floods as source of irrigation for raising the crops [4]. Locally in Pakistan the system is called as *Rodh Kohi*, derived from two Persian words *Rodh* which means riverine and *Koh* which means hilly. So in this system the rainfall received in the mountains is captured as floods in the plains. The riverine helps in the flow of water and brings it into the field. The field of spate irrigation is different from the normal field as they are larger in area and are surrounded by big embankments of 4–5 ft for retaining the water. The water once captured is allowed to infiltrate in the field till it reached the field capacity. It requires almost 2 months for infiltration of water into deep layers of soil and is used by the growing crop for the entire season. Spate irrigation is a natural system having great potential for organic cropping. As the crop are grown without the use of chemical fertilizes, herbicides and pesticides, the flood brings lots of sediments which have the natural fertility. The floods with huge magnitude usually make their way to the river, while the average and small floods are stored in the fields. The coarse size particles usually settle in the riverine and canals. While the fine silt and clay make its way to the field and build a layer of sediment greater than 50 mm per year [5, 6, 7], others have reported that it depends on the floods and may range from 1 to 50 mm year [8]. **Figure 1** shows the schematic diagram of spate irrigation.

The occupation of most of the dwellers in spate is crop farming and rearing of livestock; the farmers had the informal water user association which was previously known as the *Patidari* system, which is comprised of a *Patidar*, *Mosair* (front runner) and labourer. The main role was to labour in the watershed to divert the water by constructing stone bunds. But after the introduction of the mechanical methods of construction, the *Patidari* system was ended.

In spate irrigation after introduction of mechanization, the distribution of irrigation water and the amount of water used for agriculture have changed. The government intervention for construction of earthen bund by bulldozers provided by the Agriculture Engineering Department in Southern Khyber Pakhtunkhwa was a big assistance for the people of spate and was a step towards equal distribution, but as the agriculture engineering department was closed, the people have problems in timely construction of check dams. Now the farmers' association at village levels is involved to construct the check dam locally called as Gandi and also the other communal structures related to the spate water distribution by the tractors. But the

Figure 1.
Schematic diagram of flood. From the mountains to the fields (courtesy Mr. Nabeel Rizwan).

problems related to tractors' build structures are that they are not as strong as those constructed by bulldozers and worn away by the great magnitude of floods. Also the mechanized agriculture is useful in leveling the land and pulverizing the soils. As each year the floods bring enormous amount of silt, and the farmers capturing huge floods also add massive amount of silt which imbalances the field. Tillage has great influence on the chemical, physical and biological characteristics of soil, and it subsequently results in better plant growth and yield [9]. Also for sustainable produce from a soil, it is important to use the tillage optimally [10].

2. Historical background of the spate irrigation system

Spate irrigation is an old system being practised in various parts of the world; the most prominent countries where the system is prevailing are Eretria, Ethiopia, Pakistan, Yemen, etc. The historical perspectives and the archeological evidences show that this system may have started 2000 years ago in the Arabian Peninsula and it covered most of the Yemen area [11]. The agriculture practices and the techniques involved in spate irrigation were spread in the Muslim world through the trade and development of the countries. In Eretria the system was introduced in its eastern part by the onset of the migrants from Yemen some centuries ago [12]. The system also prevails in other parts of African countries since 100 of years. The countries include Alegria, Ethiopia, Morocco, Sudan and Tunisia [13].

The torrential floods and spate irrigation in western Pakistan are also very old, and it goes back to 330 BC [14]. The spate irrigation was an important fragment of the early civilizations, due to the economic development.

Globally the spate irrigation is practised in different continents including South and Central Asia, Middle East, North and West Africa and Latin America. It is difficult to make an accurate estimate of the land under spate irrigation, because each year the areas under spate irrigation change.

In Eritrea the spate irrigation is carried out in Sheeb area [15]. The crops are grown without the use of fertilizers as the silt brings nutrients to the field and also increases the surface level. The investigation on quantity of silt deposited and its influence on the properties of soil revealed that in the upper stream the silt was deposited in the range of 8.3–31.6 mm year^{-1}, while in middle and downstream, it was 6.0–18.0 and 5.2–8.6 mm year^{-1}, respectively. Regarding the physical-chemical characteristics of soil, it was found that siltation brings plant nutrients, but it can be further increased by the application of manures and incorporation of plant residues after the harvest [15].

Spate irrigation is a source of living to huge number of poor people. It is estimated that approximately 13 million people are directly or indirectly linked to spate irrigation around the world. This system is practised in 20 countries in different continents of the world [16]. Spate irrigation is a very old system but the work done in this area is very little. The system has achieved some attention in the last few decades as some of the organizations have intervened into the area and carried out some interventions. In Pakistan the major issue in negligence of the area was its infrastructure; in the last few years, some of the areas have been linked to the city by the construction of roads which increased the mobility of the people. The farmers are now installing tube wells through solar system, using improved varieties of crops (wheat and gram), applying tillage operations periodically, etc. [17]. In other parts of the world, the spate farmers have carried out studies on efficiency of flood water, its diversion and distribution. It has been emphasized in the publication that distribution and diversion efficiency may be achieved by proper management practices. The effectiveness of modernized package in three major countries practising spate was studied for 5 years, and it has been concluded that the spate farmers should use less number of irrigations, restrict the embankment length to less than 1 m, control overstretching the area under command, obey the rules for water right distribution to allow water to the downstream and enhance the water holding capacity of soil by conservation practices as mulch, tillage, crop residue management, etc. [16].

In Pakistan a very extensive area is under spate irrigation system (**Figure 2**). It spreads adjacently to the western mountain ranges. It starts from the southern part of Khyber Pakhtunkhwa including the Dera Ismail Khan and Tank district. In Balochistan province it is mostly practised in Sibi, Kachhi, Loralai, Gwadar, Awaran, Pishin, Turbat, Killa Saifullah, Dera Bugti, Lasbela, Panjgur, Mastung and Khuzdar districts. In Punjab it is mainly practised in Dera Ghazi Khan and Rajanpur districts and Larkana, Malir and Dadu districts in Sindh province (**Table 1**). In Pakistan the hydrological system is not only found in the northern region, but the western mountain regions also have the potential of 18.68 million acre feet (MAF) of water which flows to the plains through torrential floods [18]. It is found in various other countries. In this irrigation system, the floods after the rainy monsoon season in the catchment flow down as fast moving water in channels and reach the foothill plains. This water is used for growing crops after construction of big embanked field. The torrential floods are unpredictable both in magnitude and flowing velocity, due to which it possesses greater energy and thus creates problems for the farmers. The rainfall in the plains is very low (less than 250 mm year^{-1}), but the floods bring greater amount of water to plains. Irrigation used for agriculture purpose may be attained through rainfall and floods. As the region is resource poor and lacks modern techniques, so this result is wastage of huge quantity of water. Sometimes the situation gets even poorer as timely check dams are not constructed so the water cannot be diverted to the fields. Government and non-government organizations have intervened in the area for construction of local structure for management of floods, but still there is a huge work needed for the spate irrigation area.

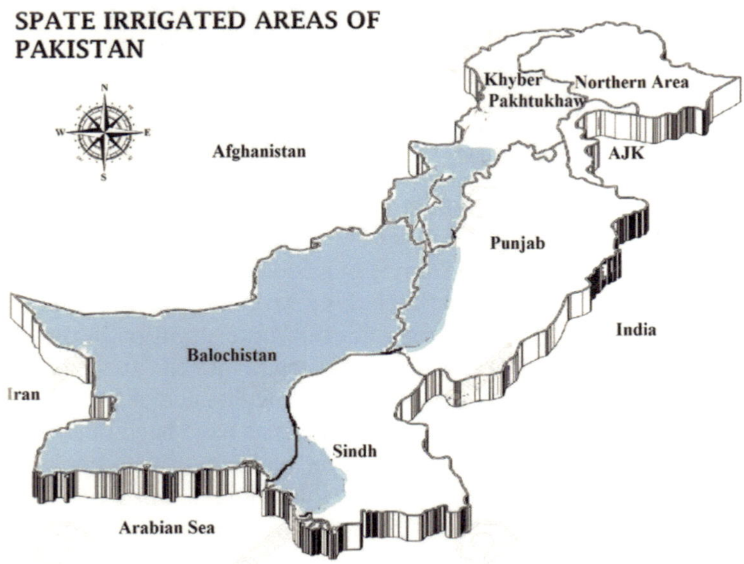

Figure 2.
Map of spate-irrigated area of Pakistan.

Province	Number of torrents	Potential area (million hectare)	Population dependent (million)	Approximate households (million)
Federal	—	0.271	—	—
Khyber Pakhtunkhwa	25	0.862	4.91	0.708
Punjab	17	0.571	2.24	0.319
Balochistan	17	4.680	5.81	0.841
Sindh	—	0.551	6.99	1.000
Pakistan	59	6.935	19.97	2.868

Sources: Agriculture Census of Pakistan, Census Organization of Pakistan, 2003; NESPAK, 1998.

Table 1.
Province wise area under spate irrigation in Pakistan.

Ref. [19] reported that there is an inadequate data available of the spate irrigation area of Pakistan to develop a proper strategy for water management. Hence, the baseline data is very vital for the sustainable development of the spate-irrigated region. To develop and plan for the future of the system depend on the availability of data or information and also proper assessment of the water resources. The changing climatic pattern has made the assessment of the available water resources more important.

3. Problem and solution of spate irrigation

The area under spate irrigation has many problems and faces the two extreme conditions, i.e. drought and floods. In Pakistan the vast area under spate irrigation is Dera Ismail Khan (D. I. Khan) and Dera Ghazi Khan (D. G. Khan) districts, the southern-most district of Khyber Pakhtunkhwa and Punjab provinces, respectively. It is adjacent to the Sulaiman ranges, the western mountain range. There are five watersheds in the Sulaiman ranges adjacent to the D. I. Khan district, which are locally called as *Zam*. These *zams* are Tank, Gomal, Sheikh Haider and Chodhwan. In districts of DI Khan and Tank, there are 27% of the total land mass (0.687 mha) under spate irrigation system only in Khyber Pakhtunkhwa, while the rest is under canal irrigation, tube well and rainfed. There is a vast network of the small and large canal known as *Rodhs* in these two districts (**Figure 3**). There are a number of problems faced by the spate farmers in this area [20]. Some of the problems and their potential solution have been illustrated in **Table 2**.

Different researchers have reported various problems of spate irrigation; flash floods brought by the torrents are of high peak but very short duration [21]. The velocity of the flood water is high due to steep gradient and greater masses, and it results in damaging the infrastructure, irrigation channel and also the standing crop [22]. The basic constraint of the torrents' flow is the conservation or management of the flood being received [23]. In D. G. Khan district of Punjab, the spate irrigation system is satisfactorily functioning for longer time, but due to continuous siltation brought by the floods, the riverine or channel has become uneven and has affected the distribution of water and has created serious management problems [24]. The capacity of the water channel has reduced due to siltation in the channels, and this has created overspill of flood water and has caused damage to the embankments and standing field crops.

Ref. [25] stated that spate irrigation using the flood water for irrigation is the cheapest technique as compared with the other methods. They have investigated

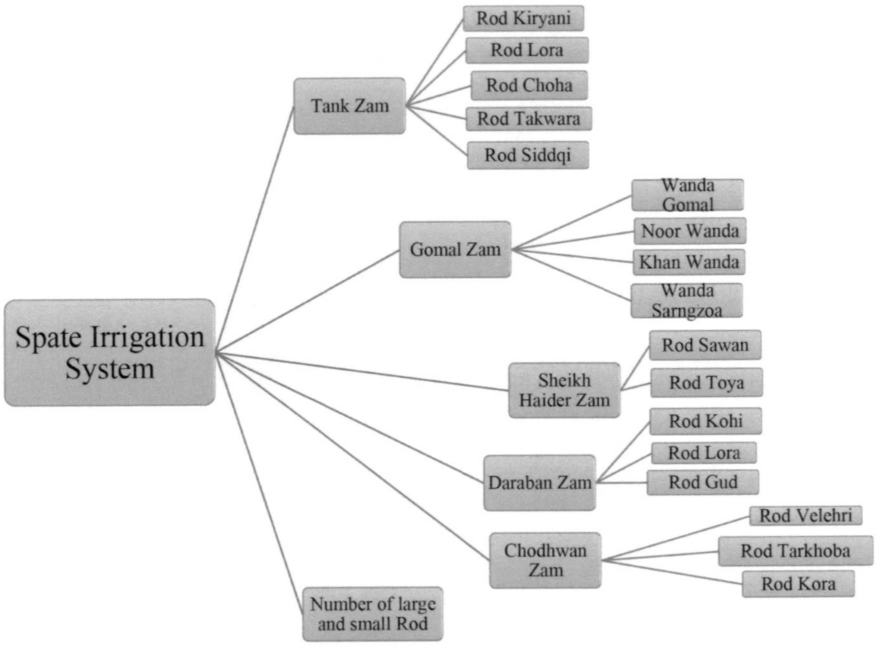

Figure 3.
The extensive system of spate irrigation of Dera Ismail Khan.

Identified problems	Potential solution
Unpredictability of the magnitude of the floods	Farmers in the spate should be prepared for the construction of check dam before time and should have fields well ready for flood harvest
Due to continuous siltation, the land is not properly leveled	The farmers should be facilitated through the government agencies in land leveling
The water distribution is uneven, and sometimes the head farmers may apply water more than once while the tail does not even receive a drop of it	The water user association functioning may help in even distribution of water between the different streams
Lack of technological development in the spate-irrigated area	The linkage between the different stakeholders is necessary for the development of area
There is less attention from the stakeholders	There is a need for the spate community to be aware to raise their voice and highlight the problems
The crops grown in the spate-irrigated area lack the access to main market	Access to market may be accomplished after the linkages are developed, and the mobility of the people enhances
The organic food produced in the spate has not gain the attention of the market or promotion	The consumers are less aware about the importance of organic products. But it is gaining importance as the health hazards and environmental problems are increasing

Source: [20].

Table 2.
The identified problems and their potential solution.

the different factors influencing the spate irrigation in Ethiopia from the year 2005; they have studied the crop choice and the requirement of irrigation. They collected data using the logit model. It was found from the study that farmers having the low cost of production, greater capital for irrigation, own labour (family) and least rainfall may take the option of spate irrigation. Also they reported that crop choice is not affected by the market. Most of the spate farmers grow cereals and pulses as their own food security. The probability of the modern techniques in the spate irrigation may increase by spending less money on cost of production, availability of water resources, variation and climatic factor.

4. Water-agriculture balance in spate irrigation

This anomalous behaviour of water opens new windows for searching water resources. We receive water either from rainfall or from floods. Both resources are the outcomes of recycling. When climate temperatures slightly rise, it causes glaciers to melt which changes the ice into water. But this change in climate is controlled and bearable. In summer when temperatures are expected to become the highest, it then dismays the environment and brings floods. On having such situations, it becomes necessary to channelize this flood into irrigation water that can be termed as water harvest. In spate areas the inhabitant or farmers receive these floods as blessing of God, and in urban area it is not more than a menace.

In spate-irrigated areas, poverty commands and people are very hard to live. They are dependent of agriculture profession as main and sole profession. Soil is hard and barren, and water table is very low and neither so link nor black top road to link between farms to market. Although the area is responsive to many crops like wheat, gram, millet, *Sorghum*, guar, melon, *Brassica* and onion, they prefer to grow wheat and gram as second choice. An irony is that farmer cannot decide, only flood decides which crop should be sown and when to sow that crop. In Pakistan, after 2010 floods, the climate sway has been observed by the scientific community. This flood was prominent for its severity; 2010 floods are notable for their severity [26]. Projections related to the impacts of climate change warn that developing countries will be the greatest victims of climate change. The Global Climate Risk Index has rated Pakistan as the seventh most vulnerable country to be affected by climate change [27]. Imbalance environment (climate change) and over- and under-watering have depicted the significant changes amongst the common strata/people. Unusual floods in terms of time and magnitude have greatly influenced the prevailing climate and also disturbed the seasonal monsoon route. Such alteration has also made upset the rainfall schedule due to which crops are being grown either early or delayed to their actual time. Thus it can be said that farmers' sowing trend has been shifted due to changed rainfall docket. An interesting thing which has been observed in locality amongst the farmer is adoption of the people according to changed climate as well as water mood/temper. That is why they apply indigenous knowledge which they have perceived from remote ancestors for the purpose of biological control, moisture conservation from the flood water, fencing the flood, diverting flood route, converting flood water into irrigation water and handling noxious weeds and baneful pests. This native cognition also covers the cultivation method, crop choice, quality seed grading, seed storage, livestock health and crop care and their safety from diseases, pest identification and eradication.

4.1 Crops

The farmers in spate are growing different crops as source of fodder, fibre and food for earning their livelihood. The farmers in the spate-irrigated area are subsistent and wait throughout the year for harvesting water and growing crops. During the dry season, they have to migrate to the canal-irrigated area to earn their living.

Mostly the cereal crops are grown on vast area [18]. These include rabi (winter) crops such as wheat, gram and kharif (summer) sorghum, millet and melon. The comparison of yield of both flood water and perennial water revealed that the yield is low in flood-irrigated area. These cereal crops are essential for the food security and survival of the spate farmers, but there is potential for certain cash crops as cotton, sesame, guar, etc. [28]. However, there is very little choice for the farmers, as it mainly depends on the amount and time of water being received. During the monsoon floods, if the water received is huge, then wheat is the option, but if it is less in magnitude, the farmers have to grow gram. The cotton crop vanished due to more insect pest and shortage of water, whereas mung bean has disappeared due to growing of other fodder crops such as millets and sorghum. The wheat and gram are the preferred crops of the area, and the farmers and the owner try hard to get enough water to cultivate gram and wheat.

The crop yield has been fluctuated by heavy rainfall or floods and limited in no rainfall or drought. Other fields responded to fodder growth with little moisture. Under spate irrigation yield of the crops like sorghum, oilseed and wheat have been reported as 745,510 and 915.5 kg ha^{-1}, respectively [18]. The spate-irrigated land has the potential to give better yield of crops (**Table 3**) if the floods are timely available and management practices, i.e. land preparation, weeds management, pest control etc., are used.

Wheat grown in area has unique quality due to which it has captured the main focus of the farmers. The peculiarities of the local landrace of wheat is that it has greater loaf volume and greater protein content, hence having a better baking quality than the improved varieties. This wheat variety is known as dual purpose wheat, i.e. both consumed by human beings and livestock, or in other words it is used for fodder during early-growing season and also as straw after harvest. The grain yield of this landrace is comparatively low but has the potential to improve if proper environmental conditions are provided. It is meritorious in many aspects, i.e. fodder, grain and even straw is utilized which is very high. The physiology of this wheat cultivar is also praiseworthy as the seed grown at a depth of 10 cm may germinate, which indicate that it has a longer coleoptile length. As most of the farming system is based on monocropping, the farmers strive for growing wheat as it is a source of food and also seed for the coming season. The landrace has the potential to grow in dry condition as well as in the absence of fertilizer. These are the main reason that wheat is the first choice subject to the availability of water or flood whatsoever.

4.1.1 Fruits

As perennial water has been mostly available to the upstream villages, there were many orchards in the spate-irrigated areas (**Table 4**). The fruit orchards

Crop	Yield (kg ha^{-1})	
	Flood irrigation	Perennial irrigation
Wheat	905	1600
Gram	500	710
Millet and sorghum	610	1050
Melons	4080	5200

Table 3.
Productivity of crops grown under non-perennial and perennial spate irrigation in D.I. Khan and tank districts of KPK (agriculture census, 2006).

Farming practice	50 years ago	Now	Reason for change
Mung bean cultivation	Cultivated on few of the area which has received a lot of flood	Not cultivated any more	Mung bean requires more water, and therefore most people turned towards gram
Cotton	Cotton was grown	Not grown anymore	Pest attack and people have changed from *Sorghum*, and millets staple food to wheat, so by growing cotton land was not available for wheat
Fruit trees	Ber (*Ziziphus* sp.) is the common fruit in all villages and date palm orchards in some of the villages as Chodhwan	Very few ber trees except for the upstream perennial water receiving villages and at household	The Kuliyat and Riwajat in which the trees are the property of landowner, and therefore the tenants did not care much for the trees
Plowing	Bullocks were used for plowing and earthen bund construction	Tractors and bulldozers are used beside the bullocks used by limited farmers	Introduction of mechanization in the 1972 onward by Agri. Engineering department and tractors became more common after provision through subsidies in 1989–1990
Tenancy pattern	The production was divided half and half with the sixth share of the owner	But now the trends of half and half share without sixth have started	To cultivate land by hired labour or temporary tenancy, the tenants do not agree for the sixth share; this is also happening in few of the area in permanent tenancy

Table 4.
Changes in spate agriculture over the 50 years.

included were date palm, grapes, mangoes, apples, ber, etc. But now most of the orchards have been turned into living places and houses; this is due to shortage and injudicious use of water and growth in population and land fragmentation. The major fruit trees found now are ber, date palm and melon. Ber is known in Pakistan as a 'Miracle Tree' as it has the potential to tolerate the severe drought condition. It is used as fruit and also a very good fodder tree; the shepherd in the spate-irrigated area buys the tree for its twigs and leaves [29]. The height of these trees may reach up to 10 ft. But now the population of this tree has been reduced due to scarcity of water, and also as the trees are the commodities of the owner of the land, the tenant is mostly not allowed to use the branches for their livestock, so the tenants do not look after them and even do not let them grow.

4.1.2 Vegetables

Different vegetables as bitter gourd, ladyfinger, tomato, onion, green chilies, etc. were grown in the upstream area and local market existed. Due to increasing population and easier access to the main markets, most of the vegetable gardens have been changed to living places. Conflicts of water usage to irrigate orchards have been also minimized. In the middle stream and downstream villages, they grow vegetables on the embankments for stabilization of the embankments with the crops for their own consumption. These vegetables include pumpkins and bitter gourd.

4.1.3 Trees

The native trees in spate-irrigated area are *Acacia nilotica*, *Tamarix aphylla*, *Capparis decidua*, *Prosopis cineraria*, etc. which are common but found standing alone in a huge field. In the early 1900s, *Prosopis juliflora* (mesquite) was introduced in by Mian Musa Alvi the servant of Dak bangle; he was provided the seed by the English officer for his barren land to control the land degradation. This tree has the ability to spread faster and has now occupied the whole spate area, and every day eight to ten lorries of the fuelwood obtained from this tree have been sent to urban area and therefore have proven to support the labourer class in their daily life. Due to the rapid spread of mesquite, other tree species have reduced.

4.2 Soils

The soil of spate-irrigated area is usually rich soil with native soil fertility, and the sediments brought about through the floods replenish the fertility status. The texture of soil ranges between silty clay and silty clay loam, and these have greater water holding capacity due to fine size and may retain the moisture for a longer period of time [30]. It has been observed that most of the floods received during the monsoon season (July–August) are being allowed to infiltrate in the soil layers for 2–3 months, and the surface is mulched with the soil during the late summer when the temperature is still high, but the water is strongly held that it does not evaporate. The same moisture is used for cultivation in late October, and if no supplementary rains are received, the moisture is enough for the whole crop season till harvest. Changing climatic pattern will have great influence on the water resources and will also affect the soil resources [31]. Climatic model showed that the direct impact of climate change on soil properties may be the losses of soil carbon which will also affect other soil characteristics like poor soil structure, soil strength, water holding capacity, nutrient cycling and increase in soil erosion due to less soil cover [32]. The soil structure is an important soil property in the spate-irrigated area as it affects the movement of water and nutrients, growth of plants, exchange of gases and activities of soil fauna. The soil structure of spate soil is usually improved by the addition of organic matter with receiving of floods. As the temperature of soil increases the rate of organic matter, decomposition will also increase, and it will destroy the structure of soil. The water holding capacity of soil is also linked with the soil structure; the soil having good soil structure has greater water holding capacity. The water holding capacity of soil in dryland areas like spate irrigated is very important as the moisture is mostly stored in soil and the plants can efficiently use the moisture. The productive soil has the capacity to provide nutrients to the plants, but the nutrients' mobility is mainly concerned with the organic matter decomposition and availability of moisture. The soil of spate has native fertility, and it replenished with each flood, but under extreme drought the soil surface becomes scarce in vegetative cover, and the following floods may result in greater soil erosion. As there is open grazing, the threats towards the erosion both due to water and wind increase. The local farmers usually grow vegetable on the embankment to stabilize the structures. As the dry spell and temperature increase, the rate of evaporation exceeds the rainfall which may lead to salinity and sodicity. Sodicity is the excess amount of exchangeable sodium accumulated on the surface of the soil. Crust formation on the surface of the soil is a problem faced by different soils around the world [33]. The crust formation is also very common in the spate-irrigated areas due to depositional and formational structure; this is due to the fact that the area of spate has the climate conducive to crust formation. The crust formation in spate-irrigated area may adversely affect the increase in the surface runoff and reduction in the rate of infiltration into the soil. The crust formation reduces the germination of seedling and eventually results in the decline of crop yield. There are different solutions proposed for overcoming the problem of crust formation such as optimal tillage operations, application of manures, crop residue incorporation, mulching, seed sowing at appropriate depth, sowing on ridges, etc. Soil in this part of Pakistan is moderately to strongly calcareous with alkaline pH. Soils have sufficient potassium content and micronutrients but are mostly deficient in nitrogen and phosphorus. The fertility status of spate-irrigated soils of Dera Ismail Khan, Pakistan, in 87 soil samples collected from different locations and analyzed for various soil physico-chemical characteristics showed that 50% of the soil were of medium texture and 45.9% were clayey fine texture. There were very little soil with coarse texture (3.5%), and saturation percentage of the soil ranged between 16.2 and 67.0%; salinity and sodicity showed that 13.8% were saline, 5.8% were sodic and 74.7% were normal soils. The fertility status of the soils under investigation revealed that all the soils were deficient in nitrogen, while phosphorus was deficient in 89.4% soils, but potassium was found adequate in 70.6% of soil samples analyzed [30].

4.3 Climate change: pros and cons

Climate change is an important term that mentions the variation in climate over time; it may be due to changes in natural events or due to human activities [26]. The climate change has greatly influenced Pakistan by frequent spells of extreme events of weather, i.e. floods, glacial lake outbursts, droughts, heat waves, etc. These extreme events have made Pakistan more vulnerable to climate change and also have killed many lives. The impact of climate change on Pakistan agriculture, economy, property, etc. is well documented [34]. The heavy flood of 2010 has alone killed 1600 people, caused damage of approximately $10 billion and flooded 38,600 square kilometres (km^2) [35].

The freshwater resources of Pakistan are very small. The primary source of water is rainfall and it provides major as monsoon rains [36]. Climate change will affect the arid and hyperarid areas of Pakistan due to uncertain and erratic rainfall pattern. The rate of evapotranspiration will increase which will also enhance the water requirement of crops by 10–30% [37].

Spate irrigation and climate change are deeply intertwined, as the system in spate is wholly solely dependent on the rainfall and floods. The time and the magnitude of rainfall also matter. Mostly the timely rainfall received in monsoon and spring seasons are crucial for the summer and winter floods which are made useful for growing agricultural crops. Also the selection of crops in spate irrigation is influenced by the time and magnitude of floods received, as there are limited options for the farmers of spate regarding selection of agriculture crops. Due to untimely rainfall either early or late, the cropping season is badly affected and may lead to a fallow season or low crop yield. The yield of wheat crop may be significantly decreased by receiving greater rainfall, but marginal enhancement in the rainfall will not be going to have any considerable results of wheat [28]. Also the livestock may be affected by the climate change in two main ways, one by the reduction of fodder and forage and second directly by the temperature intensity [38].

Other climatic constraints under spate irrigation may be increased atmospheric temperatures which may lead to higher evaporation and severe moisture stressed conditions. More extreme condition may be recorded if heavy floods are received which may wear away the big embanked field and standing crops and erode the fertile soil. In Pakistan the communities which are found to be the most vulnerable to climate change are the subsistence farmers having the small landholdings [39].

Growth of crop is highly affected by the amount of water and changes in temperature; according to an estimate, the increase in temperature by 0.5–2.5°C the agriculture productivity of the crop will decline by 8–10%, and this is estimated to be the case by year 2040 [40]. The crop growth simulation model showed that the length of growing season of major crops, i.e. wheat and rice, will decrease which will result in the decrease in yield of crop. Also the climate change projection provided by the Intergovernmental Panel on Climate Change (IPCC) states that the productivity of agriculture crop in Asia is bound to decline due to heat stress and the floods and drought tend to increase.

4.4 Water rights

Water right is the legal permission to use the water, or water laws and rights pertain to the water use by water user from different water sources, i.e. canal, floods, tube well, etc. The main aim of the water rights is to avoid the conflicts on water usage and its equal distribution, avoid the wastage, etc. Under spate irrigation there are different rules for flood water and perennial water. As the flood water has huge quantity of water, so usually distribution in quantitative term is difficult, but in the case of perennial irrigation, the distribution is quantitative. Prior to mechanical intervention, the rules in spate irrigation were strictly followed, and the government supervision was also firm.

4.4.1 Flood water

The torrential flood received in the plains is unpredictable both in time and magnitude. To effectively use the flood for agriculture and minimize the conflicts amongst the spate communities, certain rules were devised. The documented rules of water distribution are locally known as the Kuliyat and Riwajat-e-abpashi (formulae and customs for irrigation). The spate irrigation of D. I. Khan and D. G. Khan was first documented by the British Revenue administration in 1905 in the form of Riwajat and Kulyat-e-Abpashi [41]. These rules are still being followed. These rules follow the head-to-tail rule which is locally called as the saroba and paina. The earthen dams are built in the main riverine and used for diverting water. The upstream area irrigates their land first, and when the whole field is being irrigated, the earthen dam is breached. Kuliyat and Riwajat mainly emphasize on the following main points as discussed by [41]:

1. When to breach the bunds (check dams)

2. The sequence of receiving water in the canal and field

3. Rights for second and third irrigations

4. Amount of water a farmer can store in the field

The water rights rules in the spate systems of Dera Ismail Khan and Dera Ghazi Khan under the Sulaiman range in Pakistan were prepared by the revenue period during the British colonial period. The major work was done by Mr. HN Bolton during his tenure as deputy commissioner in 1908. This system was considered as an important source of tax; therefore the Revenue Administration had interest in the system. Also the water distribution right provided an opportunity to resolve the conflicts and disputes amongst the people [28]. The rules were enforced by the Rod *Kohi* department. The Kulyat Rodwar has these rules written in a register having information of all the villages, labour and also the land. Also an official was responsible for the execution of these rules, urging the farmers to rebuild the embankments and plug the eroded gullies. As according to the flood water distribution rules in Pakistan in the main riverine, the concrete structure is not allowed. Only the earthen structure is allowed; the Department of Agriculture Engineering used to provide the machinery as bulldozers, etc. for the construction of large structure which were strong enough to face the strong flood, but after the closing of agriculture engineering department, the farmers construct the structure on their own through contribution. The place for construction of structure is already demarcated; they are constructed each year on the same place. As floods are unpredictable, often the structure in the upstream is not built at proper time, so the floods make their way to the mid-stream and tail. Also the surplus water is allowed to flow to the tail after breaching of the structures. The amount or depth of irrigation to be applied is not confined; usually the farmer shows voracity in water storage, which leads to breaking of bunds, and sometimes the farmer has to remove the excess water so that the soil may come to field capacity on time for sowing of crops. In the spate irrigation rules, the farmer is allowed to irrigate for the second time if all the other fields in the area have been irrigated. But factually it is not practiced, because the influential irrigate sometimes more than two times while the tail area does not receive a drop of water. Also the number of irrigation depends on the amount of water being received after the flood. As the climatic pattern is changing, the flood-based system is also changing with severe floods in some of the years and drought in other years. The conflicts may become even worse due to water distributions with change in rainfall and flood patterns.

4.4.2 Perennial water

Perennial water is another source of water in the spate irrigation. This water is locally called as Kala Pani; it is received throughout the year in the riverine through the streams. Like the land there is ownership for irrigation water. At watershed level the water is distributed amongst the different tribes living there. If we consider the example of Chodhwan *Zam* (watershed), the largest watershed in Dera Ismail Khan District, the perennial water is distributed between the Babar and Mina Khel tribes as 7:5. These are distributed through a unit locally called *Boli*. There are 10.5 *Boli* in the share of Babar tribes, which is again distributed as 1 for orchards, 1.5 for drinking and 8 for raising crops. Each *boli* is divided amongst the *Babar* subtribe or sub-cast. These casts are Musazai, Mardanzai, Badanzai, Shakarzai, Ahmad Khel, Ibrahimzai, Safarzai and Mangalzai. Each has single *Boli*. Each *Boli* is again divided into 18–20 *wails*, and in each *wail* there are 16 *churukas*. To make it understandable, 1 *churuka* means irrigation for 45 minutes depending on the amount of water. The water rights for the perennial water are different than flood water. The water in perennial system flows throughout the year. There are a number of crops that can be grown on perennial water. The perennial water may be used for irrigation on any land depending on its availability. Usually they are grown in the areas where they do not have the flood water rights. The tenure system in the perennial irrigating land is usually 7:5 for the land and water owner and the tenant.

Climatic sway has greatly influenced the perennial water as there is less rain during some of the years which may result in less water for recharging of streams and less water for perennial. Also during the flood season, the perennial water which mixes into the flood water as the riverine for both the flood and perennial water is the same. Also the farmers in spate are turning towards digging of tube wells, and some of them have been artesian well. Due to the extensive use of these

Dera Ismail Khan				Tank	
Tehsil	**Villages**	**Tehsil**	**Villages**	**Tehsil**	**Villages**
Dera Ismail Khan	Isa Khan	Kulachi	Kot Zafar Baladasti	Tank	Kot Murtaza
	Yar Manji Khel		Naskor Nahura		Gomal
	Potah		Rori		Ghasha
	Sikandar Shumali		Mohabat		Gorazai
	Chadrhar		Kanor		Dabara
	Hassani		Hathala		Shiekh Sultan
	Hawasi		Gara Hayat		Toran Tattor
	Hattu		Kot Daulat		Kalu Prangi
	Teli		Gara Guldad		Pattar
	Faqira		Gara Sardar		Uttar
	Budh		Looni		Fateh Chadhrar
			Kulachi Gharbi		Shadda
			Kulachi Sharki		Mutta
			Gara Jana		Mian Khan
			Kot Attal		Jamal Korai
			Gara Ibrahim		Jamal Awan
			Gara Nadar Badar		Diyal
			Maddi		Daulat Khan
			Kot Zafar Firodasti		Mamrez Balock
			Saggu Gandapuri		Mamrez Pathan
			Gara Gul Mohd		Khair Awan
			Rakh Ranwal		Rakh Ranwal
					Ranwal
					Murma
					Bara Khel
					Kot Allah Dad
					Habib Wattoo
					Manjhi Khel
					Mashooqa
					Tei Malook
					Azami
					Nadir Ali Shah
					Safdar Ali Shah
					Baghwal
					Kahu
					Dagar Khan

Table 5.
Villages under Gomal Zam dam command area.

tube wells, there are chances of drying up of streams which may lead to declining of perennial water flow.

4.5 Construction of dams

The flows of spate have been utilized for irrigation for centuries, but to avoid losses and wastage of water, a huge storage dam has been constructed on one of the watersheds. Gomal Zam is the largest zam amongst the five with greater catchment area and greater command area. The Gomal Zam irrigated 24 villages of Tank and 60 villages of Dera Ismail Khan (**Table 5**). The Government of Pakistan has initiated construction of dam on the watershed in 2002. The dam is named after the name of watershed Gomal Zam Dam. The dam was completed in 2013. The storage capacity of the dam is 1.140 MAF. The dam provides as source of irrigation and operates under a well-controlled system. The command area under the dam is 65,200 ha. The discharge of water has enhanced the efficiency of water and reduced the threats of floods and drought in the command areas. It will also produce 17.4 MW electricity as Pakistan is also facing the energy crisis. The climate of the area is subtropical, hot in summer and cold in winter. The crops grown before the construction of dam were wheat, gram, sorghum and millet. Now there is a potential of growing other crops as cash, cereal and sugar crops. The perennial availability of water has changed cropping pattern, and also the land which was previously kept – fallow – was used for cultivation [42].

Gomal Zam dam is the first intervention in which the spate water has been stored in a huge dam, because the heavy flood in the spate irrigation was difficult to control and used to cause much of the damage to the property and also the lives. Due to greater mass and velocity of the floods, the kinetic energy becomes greater and the damage it causes is also massive. The dam has positive impact on the economics and agriculture of the area which are noteworthy. After shifting from spate to canal irrigation, the mindset needs to be changed. There are several projects which are running to equip the farmers of the command area with improved techniques such as laser leveling, extension activities, etc. Also the area below the command area which had the water right in the spate irrigation has become deprived of the water resource on one hand, and also the ecology of the area has also changed, as the area now has become rainfed.

5. Discussion and conclusion

The spate-irrigated area focused in this chapter has the diversified physiography, with mountains on the west and River Indus on the east. The area is distinguished in terms of climate, land and water resources. The climate varies from very hot summer temperature ranging from 35 to 45°C to cold winter with temperature of 5–15°C. The district was previously under the spate irrigation and has the landform of piedmont, with vast land. After the construction of gravity canal from the Indus River known as Chashma Right Bank canal, more than half of the area become irrigated in the downstream. The western part still remains part of the spate irrigation. With the recent development, some of the influential landowners of spate have installed tube wells to irrigate the lands. But studying the hydrology of the area, it is evident that the groundwater in the spate-irrigated area is brackish. Most of the tube wells in the area are deep installation at 400–500 ft. The construction of Gomal Zam dam has also developed the area in terms of perennial irrigation.

Acknowledgements

We would like to pay special thanks to the Flood-Based Livelihood Network, Islamabad, for their financial support for running out a research project and accessing the remote area under spate irrigation in Pakistan.

Author details

Qudrat Ullah Khan[1]* and Obaid Ullah Sayal[2]

1 Department of Soil Science, Faculty of Agriculture, Gomal University, Dera Ismail Khan, Khyber Pakhtunkhwa, Pakistan

2 Department of Plant Breeding and Genetics, Faculty of Agriculture, Gomal University, Dera Ismail Khan, Khyber Pakhtunkhwa, Pakistan

*Address all correspondence to: qudrat_baloch@yahoo.com

References

[1] Sayed SA, González PA. Flood disaster profile of Pakistan: A review. Science Journal of Public Health. 2014; **2**(3):144-149

[2] Mehari AH, Steenbergen F, Schultz B. Water right and rules, and management in spate irrigation systems in Eritrea, Yemen and Pakistan. In: Van Koppen B, Giordano M, Butterworth J, editors. Community Base Water Law and Water Resource Management in Developing Countries. UK: Cabi International; 2007. pp 114-129

[3] International Commission on Irrigation and Drainage (ICID). ICID Multilingual Technical Dictionary on Irrigation and Drainage. Revised ed. New Delhi; 2010

[4] Lawrence P, Van Steenbergen F. Improving Community Spate Irrigation. Report OD 154; HR Wallingford Limited. 2005. Available from: https://assets.publishing.service.gov.uk/media/57a08c7740f0b64974001234/R8065-OD154.pdf

[5] Tesfai M. Soil and Water Management in Spate Irrigation Systems in Eritrea. Tropical Resource Management Papers, No. 36. Wageningen University and Research Centre; 2001. pp. 1-10, 45-95, 113-184

[6] Ratsey J. Engineering in Modernized Spate Irrigation Systems. Personal Communication; 2004

[7] Kahlown MA, Hamilton JR. Sailaba irrigation practices and prospects. In: Arid Soil Research and Rehabilitation. Journal of Arid Soil Research and Rehabilitation. Vol. 10. 1996. pp. 179-191

[8] Lawrence D, Van Steenbergen F. Improving Community Spate Irrigation. Report OD 154. DFID Project R8065; HR Wallingford; 2005

[9] Rashidi M, Keshavarzpour F. Effect of different tillage methods on some physical and mechanical properties of soil in the arid lands of Iran. World Applied Sciences Journal. 2011;**14**(10): 1555-1558

[10] Jabro JD, Stevens WB, Evans RG, Iversen WM. Tillage effects on physical properties in two soils of the northern Great Plains. Applied Engineering in Agriculture. 2009;**25**(3):377-382

[11] FAO. Spate Irrigation: Proceedings of Sub-Regional Consultation on Wadi Development for Agriculture in Yemen; 6-10 September 1987; Aden: UNDP/ FAO. 1987. pp. 180-182

[12] El-Askari K. Investigating the potential for efficient water management in spate irrigation schemes using the spate management model. Journal of Applied Science. 2005;**40**: 177-192

[13] FAO. Water Managed Areas and Irrigation. FAO Information System on Water and Agriculture. 2005. Available from: www.fao.org.AG/AGL/AGLW/aquastat [Accessed: August 15, 2007]

[14] WRRI. Rod Kohi System Development and Management in Pakistan—A National Project. Islamabad: Water Resources Research Institute, National Agricultural Research Center; 2001

[15] Tesfai M, Sterk G. Sedimentation rate on spate irrigated fields in Sheeb area, eastern Eritrea. Journal of Arid Environments. 2002;**50**(1):191-203

[16] Mehari M, Steenbergen FV, Schultz B. Modernization of spate irrigated agriculture: A new approach. Irrigation and Drainage. 2010;**60**(2):163-173. DOI: 10.1002/ird.565

[17] Tadesse KB, Dinka MO. Improving traditional spate irrigation systems: A review. In: Almusaed A, editor. Landscape Architecture—The Sense of Places, Models and Applications. Rijeka, Croatia: IntechOpen; 2018. DOI: 10.5772/intechopen.71840.Intechopen

[18] Asif M, Islam-ul-Haque C. Hill torrents potentials and spate irrigation management to support agricultural strategies in Pakistan. American Journal of Agriculture and Forestry. 2014;**2**(6): 289-295

[19] Mustafa N, Ashraf A, Ahmad B, Iqbal B, Naz R. Spate irrigation potential and distribution of watersheds of rod-kohi areas of Pakistan using geoinformatics. Research Desk. 2013; **2**(4):300-316

[20] BARD. Rod Kohi Agriculture, Problem and Prospects Symposium. Peshawar: BARD; PARC and TIPAN, NWFP Agricultural University, A.R.I Dera Ismail Khan; 1990

[21] Ahmad H, Bokhari JI, Tallat Q, Siddiqui M. Flashflood Risk Assessment in Pakistan, in Pakistan Engineering Congress, 71st Annual Session Proceedings; 2011. pp. 707, 696–708

[22] Javed MY, Nadeem M, Javed F. Technical Assistance Consultant's Report on Pakistan: Additional Works for the Preparation of Hill Torrent Management Plan; 2007

[23] Saher FN, Mehdi MR. Selection of hydraulic structures for spate irrigation. In: International Congress on Irrigation and Drainage Workshop on Capacity Building and Training in the Water Sector. 2008. pp. 69-77

[24] Verheijen O, Van Aarst S. Spate Irrigation, Livelihood Improvement and Adaptation to Climate Variability and Change, International Fund for Agricultural Development. 2008. pp. 1-48

[25] Wakeyo MB, Fujimoto N. Irrigation technology and crop choices in Ethiopia: Spate vis-a-vis rainwater-harvesting irrigation technologies. African Journal of Agricultural Research. 2017;**12**(15): 1314-1325

[26] Federal Flood Commission. Annual Flood Report 2015. Islamabad: Office of the Chief Engineering Advisor and Chairman, Federal Flood Commission, Ministry of Water and Power, Government of Pakistan; 2015

[27] Kreft S, Eckstein D, Melchior I. Global Climate Risk Index 2017, Germanwatch; 2016

[28] FAO. Guideline on Spate Irrigation. Rome: Food and Agriculture Organization of United Nation; 2010

[29] Flood Base Livelihood Network. The Use of Trees and Shrubs in Spate Irrigation Areas. Practical Note Spate Irrigation. 2019. p. 38

[30] Mumtaz K, Khan MQ, Rahman S, Niamatullah M, Sadiq M, Ishaq AM. Characterization of rod Kohi soils of D.I. Khan, Pakistan. Sarhad Journal of Agriculture. 2011;**27**(4):591-594

[31] IPCC. Climate change 2007: Synthesis report. In: Pachauri RK, Reisinger A, editors. Contribution of Working Groups I, II and III to the Fourth Assessment Report of the Inter-Governmental Panel on Climate Change. Geneva, Switzerland: IPCC; 2007. p. 104

[32] Karmakar R, Das I, Dutta D, Rakshit A. Potential effects of climate change on soil properties: A review. Science International. 2016;**4**(2):51-73

[33] Nizami MI, Akhtar E. Soil crusting and management in rod kohi area of Pakistan. In: Rod Kohi Agriculture, Problem and Prospects Symposium. Peshawar: BARD; PARC and TIPAN,

NWFP Agricultural University, A.R.I Dera Ismail Khan; 1990. pp. 195-198

[34] Government of Pakistan, Ministry of Climate Change. National Climate Change Policy. Islamabad: World Bank, Asian Development Bank, 2010. Pakistan Floods 2010: Preliminary Damages and Needs Assessment Report; S. Kreft, D. Eckstein. 2013. Global Climate Risk Index 2014: Who Suffers Most from Extreme Weather Events? Weather Related Loss Events in 2012 and 1993 to 2012. Briefing Paper. Bonn: German Watch; 2012

[35] Ali A. Indus Basin Floods: Mechanism, Impacts and Management. Manila: ADB; 2013

[36] Kiran A, Qurat-ul-Ain. Climate Change: Implications for Pakistan and Way Forward. Islamabad: Institute for Strategic Studies, Research and Analysis (ISSRA), National Defence University; 2017

[37] Hussain A, Bangash R. Impact of climate change on crops' productivity across selected agro-ecological zones in Pakistan. The Pakistan Development Review. 2017;**56**(2):163-187

[38] Adams RM, Hurd BH, Lenhart S, Leary N. Effects of global climate change on agriculture: An interpretative review. Climate Research. 1998;**11**:19-30

[39] Saifullah. Climate change impact on agriculture of Pakistan—A leading agent to food security. International Journal of Environmental Sciences & Natural Resources. 2017;**6**(3). DOI: 10.19080/IJESNR.2017.06.555690

[40] Dehlavi A, Ghorst A, Groom B, Zaman F. Climate Change Adaptation in the Indus Ecoregion: A Microeconometric Study of the Determinants, Impacts, and Cost Effectiveness of Adaptation Strategies. Islamabad: World Wide Fund for Nature (WWF) Pakistan; 2015

[41] Van Steenbergen F. Water Rights and Water Distribution Rules. 2004. Available from: www.spate-irrigation.org

[42] Khan SA, Shah SA, Ullah I, Ibrahim M, Khan S. Impact of Gomal Zam dam irrigation project on agriculture and welfare of farming community in southern districts of Khyber Pakhtunkhwa-Pakistan. Asian Journal of Agriculture and Rural Development. 2017;**7**(10):212-218